BEYOND LEFT AND RIGHT

BEYOND LEFT AND RIGHT

Lies the Path to our Golden Age

Garry MacGregor

Beyond Left and Right
Copyright © 2023 by Garry MacGregor. All rights reserved.

No part of this publication may be reproduced, stored in a retrieval system or transmitted in any way by any means, electronic, mechanical, photocopy, recording or otherwise without the prior permission of the author except as provided by USA copyright law.

This book is designed to provide accurate and authoritative information with regard to the subject matter covered.

Book design by Launch My Book, Inc (www.launchmybook.com) with cover design by Erika Alyana Duran (easduran.myportfolio.com) and interior design by Booknook.biz.

Published in the United States of America by Ugly Duckling Books

ISBN
978-1-7380586-0-0 (paperback)
978-1-7380586-1-7 (ebook)

For Katie and Heather

and the horde of brilliant, dedicated, positive, creative young people like them who are leading us into a golden future

Contents

Introduction	1

Part I: The Past

Beginnings	9
The Originator	31
The Originator Breaks Free	53
The Growth of the State	69
The Inevitability of Empire	99
The American Empire	125
Life in the Machine	159

Part II: The Future

Taking Control	189
Cooperation	211
Organizing our Cooperation	221
Eve	241
Work	263
Responsible Infrastructure	281
Looking After Each Other	307
The Dominator	335
Conclusions: Eleven Things We Know	349
Notes	367
About the Author	373

Introduction

〜➤

When I became a father over thirty years ago, I tried to imagine the world that my daughters would inhabit. Which of the momentous changes now taking shape would frame their lives? What lurking problems would limit their horizons? What new options or fresh challenges would spring from promising new technologies? What was my generation handing over to theirs?

I had reason for concern. The news seemed to offer only a litany of disasters. I had friends who had chosen not to become parents in the face of a future they saw mainly as a looming series of threats—the shadow of nuclear annihilation hanging over every lurch of the Cold War, an inexorably rising population that was pushing in many ways against the natural limits of our planet, pollution tainting our air, rivers and oceans more each year, disappearing wildness with the consequent loss of habitat for many species. These problems could not be ignored. Dealing with them would not be trivial. My daughters' generation had challenges ahead of them. My generation had left much unfinished.

My hope for the future in the face of these dangers was not entirely rational. All I had was a faith in human ingenuity and the knowledge that predictions based on projections of current observations into an unknown future had always had a dismal record of success. The cleric and economist Thomas Malthus had looked around him and made dire predictions more than two hundred years ago with a reasonable analysis based on current population growth and current agricultural capabilities, but his analysis omitted the phenomenal changes in agriculture that were soon to occur in response to the looming problem. A smog event killed ten thousand people in a December week in London in 1952, leading to apocalyptic predictions about uninhabitable cities that failed to take into account the rapid phasing out of coal in favour of gas and oil that occurred in the next decade.

Many times, bleak prospects had spurred transformations that recast the range of possibilities. If you had been a Birmingham cotton worker in 1820 or an American factory worker in 1910, your grim prospects would have made a glimpse at the world that your grandchildren experienced seem entirely fanciful. On the other hand, relying on as yet un-invented transformations to solve problems is a dicey strategy: some harbingers of doom do indeed foretell collapse. The problems around me were real. They would have to be dealt with. The next eighty years during which my girls would be active promised to be both interesting and dangerous.

With the newfound seriousness that accompanied fatherhood, I felt compelled to organize my thoughts about the world. I seated myself at the kitchen table, opened a notebook, and set out to impose order on these thoughts, hopes and fears. I wanted to know how best to pull my weight in helping my girls inherit a better world. I expected this exercise to occupy a few evenings.

My first attempt at writing was exhilarating, but a few days later when I read what I'd written, I felt discouraged. I had produced a set of opinions, most of which were based either on prejudices I had never truly examined or borrowed conclusions I had too easily accepted. Comments such as why is this true? and how do you know this? soon decorated my pages in red ink. The seriousness imposed by fatherhood made me a harsh critic of my own flabby thoughts.

Once these questions were exposed, however, I found it impossible to stuff the notebook back in my drawer and forget about it. I was constantly pulled back to my writing table and to my bookshelf. My questions changed from who can we blame? to why does this happen over and over again? Why do people act the way they do? What drives human societies into such self-destructive behaviour? and what are the barriers that make it impossible for us to face certain problems and put them behind us once and for all? As I read history with these questions in my mind, I started seeing consistent scenarios beneath the changing details. Since my day job was running a bookstore, I was tasked with poring over publishers' catalogues to choose stock for the store. Books that spoke to my questions started to appear on the store's shelves. Many of them made their way home to me. Following the logical chains from one question to the next shaped my evenings. Understandings led to new questions. The girls grew older. The world continued to change. This has been background to the last thirty years of my life.

One theme that reappeared again and again was that a powerful human urge to accumulate wealth and power was at the base of many of the historical processes in our world. This lust consistently drove us into wars, into making threats and generating fears, into forcing people into poverty, into the horrors of imperialism and colonialism, and into the despoiling of common wealth for personal gain. It not only created many of the problems I saw around me but it divided us in ways that prevented us from addressing the harm. Where does this urge come from? Is this the essence of being human? Is this a flaw in humans that dooms us to our eventual destruction?

It was also clear that this was not the whole picture. There was another powerful drive, living alongside this thrust for dominance, which consistently drove people towards selfless acts. Within a family, within a community, amongst a group of friends, within a platoon in war, even between strangers who simply recognized their shared humanity, people were expressing community everywhere. This drive was often less obvious in history books and newspapers because it produced less of the bleeding and suffering that passes for news, but when you started looking for it, it was everywhere. It was another undeniable part of being human.

Are we in the thrall of this lust for power? Or are we driven by these feelings of shared humanity? Are we good creatures who have become corrupted or evil creatures who must be controlled? It was an old question, but crucial for understanding our possibilities.

I found answers in our deep history. Like every creature alive, we have been moulded by the process of evolution. It is easily accepted that our upright stance, our binocular vision, our opposable thumbs, our large brains—everything we can see—has resulted from a process of adaptation and selection. But it is every bit as true that our instincts—including those urges that drive us to act—exist because, at some point in our past, they were essential to our survival, giving those who possessed them in some measure an advantage over those who did not. After hundreds of thousands of generations, this advantage makes the drive integral to all members of all future generations. To understand the urges that drive humans today, we must look at our past.

The book starts with a recap of the evolutionary part of the human adventure. Over millions of years of change, our ancestors were forced to adapt to different environments, each making different demands on them, each inculcating different instinctual drives. Because of this variety in our journey, we contain more than one motivating drive, a confusion that can only be reconciled by the creation of traditions, taboos, laws, customs and belief systems to favour actions flowing from some drives while discouraging others.

For the last ten thousand years, our world has been shaped by feudalism, a structure which gives pride of place to the urge to dominate. Our examination of our past shows why this happened and then we follow the implications. This preference given to domination over such a long period permeates our structures, our traditions and our beliefs. It defines our problems. But it also becomes clear from this look at the past that our social structures are manmade. They are a choice. The urge to dominate does not define our horizons. It is ubiquitous only because it has been encouraged by structures we have created.

Continuing to delve into our past, we uncover a contradiction at the heart of this feudal world. A world based on competition—knights conquering their neighbours to produce larger demesnes; states conquering weaker states to create empires; companies exploiting workers and despoiling environments to amass wealth so they can grow larger and more powerful—has growth baked into all of its structures. But growth cannot continue indefinitely in a finite world. A system that must generate growth cannot survive in a world with limits. It will destroy itself. This inherent contradiction must end in grief. I had found the nub of my fears for my girls.

Luckily, the way out of this quandary is also obvious in our history. We humans choose our structures. Our evolutionary journey has given us a range of drives which can lead to a range of different ways of organizing ourselves. Basing our world on the urge to dominate is not our only option. We have structured societies around other drives in the past. History provides examples of people who successfully established taboos to hobble "the Dominator" while encouraging values of cooperation, equality and responsibility. These worlds are also natural for human beings. They will flourish when we construct settings that encourage their expression. Part II examines a set of actions that can transform our structures from encouraging domination to encouraging cooperation.

When we expand on this new option, we find that the world that emerges is one in which inequality need not define us, where war becomes impossible, where everyone is fed, where bullies are put in chains, where we happily live within the limits of our planet, and where our strengths are dedicated to providing us with leisure. We need not let the limitations of our current mindset destroy us.

The political milieu within which I made my first attempt offered me two ways forward. I could ally myself with one of the flavours on the left or I could choose one on the right. Both choices encouraged me to join the battle to win control of the machinery of growth. But both ignore the fact that growth itself is the heart of the problem. Forcing everyone to choose sides in this battle precludes any examination of the real problem. The difficulty was not that either group had the wrong answer. It was more that neither has the right question. A choice between the left wing and the right wing of the growth machine is no choice at all. It is time to get off the bird.

Part I
The Past

CHAPTER ONE

Beginnings

In the beginning, almost fourteen billion years ago, all matter, everything that would eventually become our universe, our world, all life, was compressed into an exceedingly dense point. The big bang propelled this matter in all directions. This moment of either pure energy or unimaginably dense matter is the best starting point we have. The origin of this energy or point mass is hidden from us: nothing from before the big bang has carried forward.

One second after the big bang, the temperature of this expanding universe was 10 billion degrees, much too hot for atoms to form.[1] But this plasma of electrons, protons and other subatomic building blocks cooled as it expanded. One hundred seconds later, the temperature had dropped by a factor of ten and hydrogen and helium atoms could start to come together. As temperatures continued to drop, these simple atoms began to fuse into heavier elements. Because this process was uneven, lumps occurred, and gravity took a hand in forging progressively larger bodies. Billions of years of this resulted in a universe of stars and planets. A desolate rocky earth realized its present size, shape and solar orbit about four and a half billion years ago.

Stars continued to fuse their constituent hydrogen and helium, in the process becoming radiant sources of heat and light. One such was our sun, the focus for the orbits of several other planets as well as our earth. The interior of the earth was also active, generating sufficient heat to keep the core molten. This powered frequent volcanoes which created a gaseous atmosphere of methane, ammonia, water vapour and neon around the planet. This envelope trapped heat and temperatures rose into a range where liquid water began to accumulate.

And then life appeared. Fossils of bacteria have been found[2] dating from three and a half billion years ago. This change from dead combinations of atoms to something capable of sustaining and reproducing itself was a leap. Perhaps the possibility of life always existed, lurking in the plasma from before the big bang. Perhaps there is a yet undiscovered process whereby collections of atoms, under the influence of light, heat and water, can become alive. Whatever spark got us started, at that point we were launched onto an exciting evolutionary adventure.

These first living creatures consumed carbon dioxide, using the carbon and exhaling the oxygen which was toxic to them. This changed the world around them. In a scant one and a half billion years of life, the earth accumulated an oxygen rich atmosphere. Life became progressively more diverse. Those better able to use a food source, less likely to become food for others, or hardier in some aspect, increased in numbers at the expense of the rest. Mutations or gene transfer occasionally created new variants. The fossil record shows a steady increase in both variety and complexity. Over hundreds of millions of years, billions upon billions of bacterial generations, the niches that could support life filled up.

1. Beginnings

If random mutations were the sole driver of diversification, then things would have tended towards stability. Each member of each species would have become optimally adapted to its niche. New variants would face such insurmountable competition from the start that they would never attain a foothold.

The story of evolution, however, is not steady diversification culminating in a stable system. It is a history of violent events such as volcanoes, earthquakes, asteroid impacts and ice ages overwhelming environments and causing extinctions, empty niches and new starts. Such events were frequent, geologically speaking. As astronomer Fred Hoyle writes:

> ... about 5000 giant meteorites with diameters of more than a kilometer have hit the earth over the past 600 million years, with an average strike rate of one per 120,000 years. Meteorites with diameters greater than 300 meters must hit the earth once in every 10,000 years ... A giant meteorite is ... capable of spraying up into the stratosphere very much more debris than a volcano can, at any rate the volcanoes of which we have experience.[3]

Debris from such meteorite strikes can shade the earth from a significant portion of the sun's heat for at least the decade it would take for the dust to settle. The loss of this heat would be catastrophic for many species. The fossil record does indeed display discontinuities where many species suddenly and simultaneously went extinct. Such an event about sixty-five million years ago is fingered as the probable cause of the extinction of the dinosaurs. That mass extinction is but one of many in our past.

The happy side of this for some, however, is that those species lucky enough to have a few of their numbers survive the crisis would inherit a world with a greatly reduced number of competitors and predators. Previously occupied niches would be available for colonization. Species moving into these niches would then be conditioned by these new opportunities. Without periodic catastrophes, evolution would have progressed towards a stable system and stopped. It was the catastrophes that powered the process along by regularly delivering new opportunities.

Catastrophes also powered the occasional leap from one species to another, a process requiring the simultaneous change in a group of characteristics. Such leaps are also numerous in the fossil record, one species following another with no record of intermediate variants. The process would work as follows: in a crisis, all members of a species would be plunged into a world that differed in significant ways from the one that had shaped their bodies and honed their abilities. In most cases that meant death for the individuals and extinction for the species. But occasionally a species would have a group of characteristics or skills which, when used differently in the new environment, would allow a few of their members to survive. These survivors would be atypical—the top percentile in strength, in height, in tininess, in speed or some other deviation from the norm that would help them fit into their new situation. This initial radical culling would prune the gene pool in ways that emphasized these atypical features. Powerful selection pressures would then continue to act on the survivors forcing further modifications in the group of characteristics they would need to thrive in this new environment. When a few of them had established themselves, their numbers would start to increase, and a new species would have appeared. Such change occurs when a group is driven to the very brink of extinction but, instead of succumbing, passes through the crisis into a new niche they can grow into.

Micro-organisms spread throughout the seas. At each crisis, although many were eliminated, there were always survivors to grow into the relinquished niches. And occasionally, a new species would appear. After a few billion years of this slow elaboration, around 590 million years ago, an explosion of multicellular life occurred in the seas. Multicellular construction allowed increasing complexity including the first fishes. Around 435 million years ago, the first plants appeared, first along the seacoast and then inexorably spreading inland. As these plants also consumed carbon dioxide while emitting oxygen, their success further oxygenated the atmosphere. This was starting to become interesting: a temperature where water is neither steam nor ice, abundant water, an oxygen rich atmosphere, and a world becoming progressively greener.

One dying species of fish, attempting to survive in the drying margins of some evaporating sea, repurposed a bladder that had given buoyancy to take its oxygen directly from the air, a piscine snout rising gasping out of the swamp. That allowed them to flop from puddle to puddle as the swamp dried out. As Loren Eiseley described these first steps: "It was not the magnificent march through the breakers and up the cliffs that we fondly imagine. It was a stealthy advance made in suffocation and in terror amidst the leaching bite of chemical discomfort. It was made by the failures of the sea."[4]

From that fish the reptiles descended, large creatures who dominated the land from 200 to 65 million years ago. Plants also continued to colonize inland, still having to rely on spores and rhizomes to propagate. But near the end of the reptiles' tenure, closer to us than 100 million years ago, a new type of plant appeared that could reproduce by generating an embryonic plant able to wait, travel and sprout when conditions became propitious—a seed. Flowers soon followed, and grasses, and of course, seeds, fruit and nuts. These new sources of food soon conditioned a range of animals to take advantage of them.

Sixty-five million years ago, a large meteorite struck the Caribbean, filling the atmosphere with debris, generating worldwide fires, tidal waves, blockage of sunlight and other mayhem. Many large reptiles, such as the dinosaurs, failed to survive. Mammals, however, had recently appeared on the scene in response to the new seeds, nuts and grasses and some of these survived the crisis, probably because of their small size and because seeds and nuts can remain edible for the time it would take to ride out the crisis. When the dust had settled—and I mean that literally—mammals found themselves in a land where there were few large predators and plenty of food. They flourished, moving into the many niches that had been left unoccupied. The fossil record from this time shows the sudden advent of a range of new mammal species including the first primate, a four-legged, clawed creature about the size of a rat.

The primates were outcompeted in the richest food areas, the meadows, by the ancestors of animals like the chipmunks and prairie dogs. They retreated into the forests, areas of sparser food but greater protection. Trees provided escape from ground-based predators as well as a variety of seeds, fruit and nuts to eat. However, this new habitat made demands on them. To live safely in the trees, they had to develop an opposing digit on their front feet so they could grasp branches as they climbed. They needed to be able to gauge depth as they leaped from branch to branch, so both eyes had to be able to focus simultaneously on the same object. After incorporating these adaptations as well as others, their survival became secure. By fifty million years ago, a group of small arboreal primates was living successfully in the forest canopy. Their descendants moved into other niches in the forest.

At some point before ten million years ago, one of these species—the common ancestor of gorillas, chimpanzees, bonobos and humans—had established themselves successfully in central Africa. As they thrived, they spread into different habitats. In the next crisis, separated groups of them were forced along different lines of development. This happened first with the gorillas who appear in the fossil record as a separate species about ten million years ago. Then, about eight million years ago, the animal that would develop into the human species diverged from the one that would eventually give rise to both chimpanzees and bonobos. We know this from comparisons of the DNA of the current members of these groups.

We don't know the details of how these divergences were forced. But one scenario put forward by the marine biologist Sir Alister Hardy illustrates how this could have happened.[5] I will follow it here. If not proven to be accurate in all particulars, then some similar series of events involving isolation followed by environmental change must have occurred.

The scenario starts with a small group of this ancestral primate living in northeastern Africa just before eight million years ago. This region around the Horn of Africa is continuously being reshaped because the tectonic plates on which the land rests are pulling apart at a rate of from 1 to 5 mm per year.[6] This results in frequent volcanoes and lava flows, and the formation of sheer escarpments surrounding a deep depression holding a series of major north-south lakes—the Rift Valley system. These valleys could isolate a forest dwelling group living on the east side from their fellows who remained on the west.

Then the climate changed: forests in this region were replaced with savannah. Those on the west side migrated towards the Congo Basin, able to find new forested homes on either side of that river, eventually becoming chimpanzees on the south and bonobos on the north. On the eastern side of the rift, however, the only areas that remained forested were pockets at higher elevation and they too continued to diminish in size. Our ancestors retreated into those shrinking enclaves. One such area was the Danakil Hills, a highland on the tip of the Afar Peninsula, right at the southern entrance to the Red Sea. When the next earthquake allowed water to flood onto the low lying Afar Plain, the Danakil Hills were transformed from a peninsula into an island. As the climate continued to warm and forests at higher elevations disappeared completely, our group found themselves trapped in a world without forests. They could no longer be the arboreal creatures they had evolved to be.

There may have been other groups trapped in other enclaves on the eastern side. They may have died off immediately or survived to succumb at some later date. Their numbers were so small that no evidence of what happened has been found. What we do know is that all early human fossils so far discovered have been found on the east side of the Rift Valley; chimpanzee remains have only been found on the west. We have evidence of earthquake activity at that time. We know that the climate warmed and the forest cover disappeared. We know that the Afar Plain was flooded at that time.

The sea that covered the Afar Plain was shallow and warm. Though it was a barrier to travel, it could provide both food and protection if these creatures could adapt to use it effectively. A new survival strategy emerged which involved harvesting food from the sea and seeking relief from the heat by immersing themselves in the water. At night, they could sleep on shore near a source of fresh water. At times of the day or the year when the heat was less intense, they could gather plants from the grasslands that had replaced the forests. If they could master these changes, they had a route to survival. But this was not a trivial accomplishment. It demanded fundamental changes to both their bodies and how they lived their lives.

A body covered with hair is a good insulator in air and offers protection against the sun, but it weighs you down and loses its thermal insulation when wet. Other mammals that have moved from land to sea, such as whales and dolphins, have abandoned fur in favour of managing their insulation needs with subcutaneous body fat. These primates took the first steps along this route, shedding fur and increasing their surface layer of fat. Even today, hundreds of thousands of years after moving decisively back onto land, we have higher levels of subcutaneous fat than do all the other primates. A healthy body fat percentage for a human ranges from between 14 to 31 percent in comparison to the 9 percent body fat found in other primates.[7] Our ancestors shed their body hair because it was a hindrance in the water. They became a naked ape.

Moving around on all fours or reverting to knuckle walking was neither comfortable nor effective in water. Being mouth breathers, it was important to hold their heads above the waves. There was a strong incentive to move about in an upright stance. This change was facilitated by the buoyancy provided by salt water. Over time, selection favoured the changes that made an upright posture a natural stance. They became an upright naked ape.

They increased in size. Animals that spend a lot of time in water tend to grow larger because of the buoyancy and because seafood rich in protein can be harvested on a year-round basis.

There also had to be changes in the ways they interacted with each other. These primates were already social beings. Gorillas, chimpanzees, bonobos and baboons—their close relatives—all rely on a troop to deliver protection, comfort, meaning and food. None of them lived as isolated individuals. All of them, including the bonobo with their unique ways to blunt expressions of power, rely on a dominance hierarchy even though the details vary from species to species. Our ancestors would similarly have lived in a dominance hierarchy of some form before being stranded in their new circumstances.

The details of this organizational structure reflected their daily activities, chiefly the hunt for food and vigilance against predators. The practices that supported this structure, whether they be mock battles between putative dominant males, bullying of lower ranking individuals, or grooming sessions to cement alliances, had to occur safely during their daily travels and activities. But, since their situation had changed, activities that supported the old dominance hierarchy were no longer a part of their lives. Safety had a different face: large predators had migrated off as the climate changed and there were none on the island. Gathering food like clams or seaweed were solitary pursuits, unlike the group hunting parties or excursions into dangerous territory to pick fruit safely. While their bodies were being reinvented by the unrelenting selection pressures, their social structures were also being reshaped by their new habits.

Any new social structure had to address their three major needs—food, safety and the raising of young. No significant predators meant that they did not have to submit to a dominant male in a hope that he would protect them. There was little need for a rigid structure to coordinate food procurement; food could be harvested individually as one wished. Because the group was not forced to remain tightly together for safety, individuals could wander as they wished, so a dominant male had no assurance that he was the father of new births. Domination became impossible to impose and it no longer had benefits. For the first time in primate history, a group could meet their needs safely without having to submit to a dominant male. This was revolutionary.

Some indirect fossil evidence does exist for the dropping of a dominance hierarchy at this point. A universal difference between fossils from our ancestral line and that of all other primates is in the teeth. All primates except humans have large, pointed canines. When an ape is angry and he curls back his lips to express this, it is these teeth that make him look ferocious. Indeed, producing a frightening appearance is one significant purpose of these teeth. If easier eating or hunting were their major function, such teeth would be as prominent in females as they are in males and this is not always the case. Also, large canines exist in species with very different diets. The importance of these teeth lay in the displays of ferocity and in the lethal effectiveness in the ensuing fights that determine the dominance rankings within a troop. Fossils of our ancestral line show small canines roughly the same size as the other teeth. This implies that dominance was abandoned at this point in favour of a method of organization that did not rely on displays of ferocity. In fact, large canines were so successfully removed from our gene pool at this time that we should suspect some culling process whereby individuals with such teeth were removed, perhaps through banishment or death.

The idea that dominance was shed is further reinforced when we look at tribal cultures. Though there were hundreds of thousands of generations between this small band on Danakil Island and all current tribal cultures, we would be wise to consider that features shared by all tribes, widely spread geographically, would have a long history. And, as the anthropologist E. R. Service summed up:

> Hunting-gathering bands differ more completely from the apes in this matter of dominance than do any other kinds of human society. There is no pecking order based on physical dominance at all, nor is there any superior-inferior ordering based on other sources of power such as wealth, hereditary classes, military or political office. The only constant supremacy of any kind is that of a person of greater age and wisdom who might lead a ceremony. Even when individuals possess greater status or prestige than others, the manifestations of the high status and prerogatives are the opposite of ape-like dominance. Generosity and modesty are required of persons of high status in primitive society and the rewards they receive are merely the love or attentiveness of others.[8]

Of course, adopting a new organizational strategy would not have been a conscious choice. There were no decisions involved, no committees or feasibility studies. They were trapped in the chaos of a process that they were powerless to direct, driven by their needs in the face of social collapse. The new form that would grow to replace the dominance hierarchy would have to reflect their present needs, created by their new day-by-day activities.

They did, however, have one major problem which required a social organization to solve. Because their impulses had been conditioned by millions of years in a dominance hierarchy, they all still felt powerful drives to dominate and to react to acts of domination. The stronger males would still strive to assert their precedence by threatening others. And when conflict occurred, as it surely would, it was now harder to end with a gesture of submission indicating that the loser accepted the winner's superior place in the hierarchy. These gestures had become meaningless because the loser could just move out of sight and continue to do whatever he wanted. Gestures of submission had previously been able to turn much conflict into display. Now that they were meaningless, every scuffle risked becoming a battle to the death. Their biggest need was for a structure that would make these battles impossible or, at the very least, less dangerous.

Because they were confined to an island, they could not split into subgroups and move apart if conflict occurred. They could not banish troublesome individuals—they were stuck with each other. Eruptions of violence within the group were sure to inflict regular injuries and deaths. Being a species with widely spaced single births, such an increase in the death rate would doom their hopes for survival. The upright hairless ape was in danger of disappearing at its own hand if it could not control the violent urges within the group.

The logical route to protecting oneself from a violent individual was to gather a group of mates and, acting together, make him back off. A few individuals of middling strength would have been able to discourage an attack from one strong individual. Such banding together would not have been a leap for them as grooming alliances within dominance hierarchies are seen in chimpanzee and bonobo troops today and would have been a regular and important part of their former lives. At some point, a group of individuals would have given believable signals that an attack on one would be met by a joint and unforgiving response by them all acting together. These early groups would probably not have included the strongest individuals because the strongest were most likely to be the problem. They would probably not have included the weakest because they would be of little help in a battle. However, these alliances of middling-strength individuals, when they proved to be successful, would grow as others wanted to join so they too could gain protection. These newcomers would be welcomed because more members would permit a stronger, quicker response. As these alliances came to include most members, a new social structure began to emerge. When the alliance included everyone, the violent episodes could be made to disappear. These primates survived by producing an enforceable taboo against internal violence. Taboos on violence within the group are a vital part of all human societies. This is where it began.

This new organizational strategy was not easy to establish, flying as it did in the face of a very powerful drive. Coalitions were prone to break down as any member's own urge to dominate could cause nascent groups to disintegrate. Individual members were always vulnerable when not close to their fellows. But, the unrelenting fear of the Dominator always forced new coalitions to follow every failure until internal violence finally became controlled. Groups permanently removed intransigent bullies, probably by mob executions. Everyone worked to suppress their own violent urges lest they be perceived as a danger to the group and become the focus of the mob. As the taboos became controlling, the need for mob enforcement became rare. An expectation of cooperating in a group replaced the dominance hierarchy. This, in turn, significantly altered the individuals themselves.

Regularly removing the most aggressive in favour of those who can most successfully cooperate exerts selection pressures on the species. These early humans made themselves into new creatures when they developed this new social structure. There are a relevant series of experiments that illuminates the nature of this change.

Dmitri Belyayev, a Soviet geneticist, was hoping to understand the early domestication of wolves into dogs. Starting in the 1950s, he decided to selectively breed for tamer silver foxes, an animal available in large numbers on fur farms and previously known for not adapting easily to either captivity or humans. He chose 200 of the least frightened foxes—initially the ones that snarled least when their cage was opened—and separated them, making them the parents of the next generation. No special training or contact beyond what was given to all foxes ensued, just selection each year for the least aggressive individuals. By the fourth generation, the pups were approaching people and starting to wag their tails. By the sixth to tenth generation, they could be tamed. Surprisingly, and of particular relevance to our discussion, other changes occurred along with this change in disposition. He noted physical changes such as curlier tails, floppier ears, slimmer heads and more variable colouration, often with white patches, as well as behavioural changes which included greater impulse control and breeding more frequently and at younger ages. After more study to understand what was happening, the eventual conclusion, as summarized by science writer Matt Ridley, was that:

> ... in selecting for docility, ... [Belyayev] had unwittingly promoted a delay in the migration of the animals 'neural crest' cells during development ... Richard Wrangham, an anthropologist at Harvard University, hypothesizes that neural crest cells are also crucial to those parts of the brain that regulate stress, fear, and aggression.[9]

With the silver foxes, choosing for greater docility meant choosing for those individuals who experienced an important stage of foetal development later than the norm. Repeatedly selecting for this characteristic had continued to extend this delay. This produced a range of changes other than friendliness.

For our isolated troop of upright, hairless apes, creating alliances of middling-strength individuals enhanced the survival rate of those who could most easily cooperate with others while excluding those who were incapable of doing so. This selection process not only provided a new form of social organization, it hastened the bodily changes that made us more distinctly human. It led to greater impulse control and lowered the level of stress we feel when functioning in groups. This then eased the formation of later alliances—a powerful feedback loop. In a very real sense, we tamed ourselves every bit as much as Dimitri Belyayev tamed his silver foxes.

After what would have been a perilous period surviving on the edge of extinction, forced to adapt anatomical changes and adopt new habits, this group finally became secure. Whether it took thousands or hundreds of thousands of generations, they found themselves on a coastline rich in seafood, moving in an upright posture, organized in a troop that could establish and enforce taboos against internal violence. This transformation offered significant opportunities.

Hands and arms no longer required the strength necessary to swing through the trees. As they gave up strength, they developed increasingly precise movements of the hands and fingers. Other primates make and use tools, but when our ancestors became capable of more precise movements, they could call on tools to play a more important part in their lives.

Spending long periods in water would have impaired their traditional methods of communication. As Elaine Morgan said:

> When you are swimming you can't draw yourself up stiff-legged; you can't make a swift controlled forward dash for two yards and then stop dead; you can't maintain unflinching eye contact with your antagonist with the odd wave sloshing over your head or an undertow hauling you backward; you can't appease him by presenting or dominate him by mounting; you can't loom over him at your full bipedal height; you can't humiliate him with a 'cut-off' ... Gradually, over many generations, it became borne in on the hominid, ... that the only time he got his reward while in the water was not when he bristled or scowled or swung his arms, but when the noise came out of his throat, a phenomenon in himself which he'd never paid much conscious attention to, although he'd recognized the noises when they came out of his companions. ... He had the incentive to work hard on bringing this physiological function under his conscious control.[10]

Living in cooperative groups also increased their need to communicate in organizing their activities.

Over time, they relied more and more on both tools and language. This selected for intelligence because troops containing the greater number of more intelligent members were better at finding those particular rocks that could be used as tools, to envisioning and fashioning other tools and to communicating more effectively. Brain size and mental capabilities began to creep up.

However, greater intelligence is a feature that can help most animals in most survival strategies. Why did it occur in these creatures and not in the other primates who also communicated and used tools in rudimentary ways or, indeed, in any other animal at all? As it happens, countering the obvious advantage of larger more active brains, there is a cost because brains are very energy intensive. In adult humans, our brains consume 20 percent of the energy of the body at rest, as opposed to only 9 percent in the average chimpanzee. Large-brained members would have the advantage over their smaller-brained fellows day after day, year after year ... until the next food shortage. Then they would be the first to die because of their need for more food energy in order to keep functioning. Periods of food shortage are a certainty in the lives of most species and having every individual of a certain type be the first to succumb is a sure way of removing that tendency from the gene pool.

Diet is the crucial factor in allowing an evolutionary drift towards a larger brain. As Elaine Morgan summarized:

> ... a seashore diet could have had special advantages in relation to brain growth in addition to those of abundance, variety and year-round availability. ... the building of brain tissue needs a consistent one-to-one balance between Omega-6 and Omega-3 fatty acids. ... The latter type ... are relatively scarce in the land food chain, but predominate in the marine food chain.[11]

Having the coastline as a gathering area made the larger brain possible. When selection for intelligence occurred, as it probably did with every animal, it could proceed unchecked because of year round availability of marine sources of food rich in omega-3 fatty acids. This diet caused changes in cells throughout the body, including the brain. The selection process promoted more brain cells, new types of brain cells, more neural connections, and the larger skull needed to accommodate this new organ.

Here our group would have run headfirst into another problem. As the skull got larger, it created difficulties in the birthing process which imperiled the lives of both babies and mothers. Those giving birth earlier would be the most likely to survive because, at this early stage of development, the baby's skull was smaller and more flexible. Early births became a general trait which selection continue to advance. The downside of this was that early births necessitated an increasingly long period of infant helplessness. Luckily, the lack of predators and the ability of the group to cooperate made this possible. Also, the ligaments in mothers' hips adapted so that the size of the birth canal could increase during birth. For thousands of generations, there would have been a struggle between forces selecting for larger brains and high mortality rates during birth. The change from ape to human was not an easy one.

Giving birth to helpless young forced further elaborations in their evolving social structure. The whole group needed to cooperate to assure that infants were cared for and protected until they were competent. This task would spark greater use of language, adding to the push for larger brain size. Increasingly helpless young at birth, earlier births, the tribal cohesion necessary to care for helpless children, the growing need for more extensive language, and larger more competent brains were all advancing at the same time, each stimulating compensating changes in the others.

These primates reinvented themselves to survive. They became hairless. They stood upright. They advanced in the use of tools. They developed the rudiments of language. They developed a more competent brain which allowed them to make greater use of these other innovations. And they became tribal beings, each curbing their own impulses in order to conform to accepted taboos on violence.

The evidence from this period is incomplete so specific details may have to be updated as new evidence presents. However, there is much that we know for certain. We know from genetic analysis that we, the chimpanzees and the bonobos had a common ancestor about eight million years ago. We know from geology that earthquakes created the Rift Valley and turned the Danakil Peninsula into an island. We know that climate warmed at this time. We know that chimpanzee and bonobo fossils after this period exist on one side of the Rift Valley while fossils of human ancestors are found on the other. We know from comparisons with the other primates that we became upright, hairless and larger with an increased level of subcutaneous fat and a reduction in the prominence of our canines. We know that all hominids exist in tight groups and that all historic tribes have inflexible taboos on internal violence. Our story incorporates and explains these facts.

In this quick tour through our past, we uncovered the roots of two powerful impulses driving human behaviour. Our history in hierarchies—as mammals and primates—has inculcated in each of us an instinctive urge to dominate others and to claim benefits for ourselves with strength. We are the Dominator. This aspect of ourselves can be controlled with taboos but can never be eradicated. It is a fundamental part of the animal inheritance of each person.

And secondly, after we tamed ourselves and moved beyond the dominance hierarchy, we lived for hundreds of thousands of generations with a driving need to be embedded within an encompassing group. Our very existence depended on basing our lives within a larger group of others. We became the Cooperator. This too is a crucial part of each human being.

When we became the Cooperator, the world divided between those within our group to whom the taboos apply and those on the outside who are seen first as threats to our collective safety. Also, the vital importance of the cohesion of the group urges us to close ranks and expel any who may be flouting the norms lest their individuality become a threat to solidarity. The taming process made us creatures who are powerful as a group, but it also made us lethal to the "other."

And then the tectonic plates shifted again. Further earthquake activity created a bridge back to the continent. Travel to and from the island became possible. And that blew the hard-won stability they had just created all to hell.

CHAPTER TWO

The Originator

~~>

For three or four million years, the troop had been isolated on their island, protected from the rest of the world by a warm shallow sea. During most of this period, they would likely have numbered less than a few thousand individuals, possibly even at times as few as a couple of hundred, constantly facing the possibility of extinction from disease, from violent deaths when the nascent taboos failed, or from too many deaths in childbirth as skull size crept larger. They would have spent days at the water's edge finding and consuming food, playing, socializing and swimming to avoid the worst of the heat. Food was plentiful and high in protein. Predators were absent.

Fear of the alpha males had induced groups of allies to remain constantly together. Others joined these zones of safety, grooming together, acting as lookouts and sharing food. Since these activities decreased the risk for everyone, the initial groups had an incentive to accept all newcomers. As the groups became larger and more permanent, they became the base for meeting more of their needs. Child-rearing, for example, was more successful in such an extended family. Increasingly, the alpha males found themselves unwelcome in any group, denied everything from mating partners to food sources. They too had to reluctantly accept the new norms or be destroyed. In time, everyone was in one of these groups of a few dozen people living along the water's edge. The taboo against violence within the group became universal. Life became secure.

As they met more of their needs as a group, more aspects of their lives became specified, especially in any area where conflict could potentially occur. For example, it was important that recognized rules would regulate the competition for mates. Such rules had to point out the partners with whom sexual activity was and was not permitted for every individual. It was not important whether the units so defined were monogamous, polygamous, or polyandrous, arranged by the individuals involved or by the community, based on affection, tribal choice or a lottery. The important fact was that everyone understand the boundaries so that battles for mates would be rare and that there would be a recognized way of dealing with such conflicts if they did occur. The emerging tribal structure had to incorporate some definition of a family unit. Similarly, for other areas in which the threat of violence had previously enforced order, such as the sharing of food, new traditions evolved that replaced fear of the strongest with a need for tribal cohesion. The importance of a zone of safety led inexorably to a set of rules, traditions and taboos. Fear of the Dominator within themselves and each other forced these people to invent society.

As this arrangement proved to be effective, and as anatomical changes reduced mortality in childbirth, the population could finally begin to grow. Since the intimate proximity of the individuals in the coalitions was essential to their effectiveness, groups had to remain small enough to allow constant togetherness. Thus, a growing population led to the island becoming populated by an increasing number of small connected tribal units. Regular contact encouraged choosing mates from other tribes which kept many of these groups related. As long as food was plentiful, confederating was natural. But distance and unfamiliarity allowed distrust to grow. As their numbers grew, the "other" crept into their lives.

When a route was suddenly opened to the mainland, most would have been unlikely to see this as an opportunity because moving would mean abandoning their predator-free zone and secure food sources. But we know that occasionally some did venture out. Probably, the reason for this lay in the human propensity for organized violence against the "other." Small groups may have fled in fear during breakdowns when the taboos failed. Alternatively, some may have been exiled after breaking the taboos. Or perhaps weaker troops were driven off the island in confrontations over resources when the population did start to increase.

When they left the island, they had a few options. They could have stayed close, settling on the continental side of the sea. This would have offered access to familiar food, though at a cost of exposure to predators. Further migration from the island could have gone three ways—north along the Red Sea, south into the Horn of Africa, or inland down the Afar River and the extensive lake system which offered a watery avenue into the interior of the continent. Deserts in the north would be a challenge. Going south, they would run into open coastline, changing the available seafood. Neither of these areas were good for the deposition and discovery of fossils, so people may very well have moved into these areas with us being none the wiser. The river and lake system also offered challenges. They would be leaving saltwater food sources behind while still having to deal with predators. The advantage of the inland route for us (if not necessarily for them) is that it included bogs and swamps where the occasional body could drown, sink and end up fossilized in a region where erosion has exposed some of these fossils to our eyes millions of years later. We know from the fossil evidence that some did head inland and survive. A river or a lake provided some protection for a swimmer. It offered fresh water to drink and seafood to eat.

That inland route includes the Awash Valley and the Olduvai Gorge, river valleys whose cliffs display sedimentary layers from the floor of this ancient lake system. The odds are steep against finding a human fossil from millions of years ago anywhere but this is one of the few areas in the world where a search for such fossils can even be contemplated. Consequently, this area has become popular with fossil hunters. Their activity has been rewarded with finds that are variously dated from about four million to about one million years ago with the fossilized bones becoming progressively more human over that period. The fact that fossils have not been found elsewhere, for instance north or south along the Red Sea, or at Danakil itself, speaks more to the way that deaths would have occurred in those spots and to the subsequent patterns of erosion than to necessarily any greater or lesser density of hominids. Palaeontologists continually stress the adage that an absence of evidence can in no way constitute evidence of absence, however much we may desire to arrange our few facts into a complete narrative.

One of the oldest fossils found was dubbed "Lucy" by her finder, Donald Johanson. Sometime around 3.2 million years ago, a short upright female died near Hadar in the Awash Valley in present day Ethiopia in a way that allowed her almost complete skeleton to become fossilized. This lay buried until the eroding surface reached it last century.

Luckily, someone who recognized what they were seeing found her in that short period after the fossil had begun to be exposed but before it had been scattered by further erosion. Her troop may have just arrived at the time of her death, they may have been just passing through, or they may have migrated in well before then and set up home. She may have been typical or an extreme outlier. Since fossils of this age are so rare, we do not have enough evidence to answer any of these questions. There may also be older fossils in this area that have yet to come to light, making her not part of the first wave, but a later incomer. This find, however, does tell us that, after three to four million years of isolation on the island, at least one upright walker, standing about three to four feet tall, with small canines, and a brain case that had not yet grown much larger than that of chimpanzees, was walking on the continental side of the sea.

Several hominid fossil fragments with a different set of characteristics were found by Louis and Mary Leakey further south along the Rift Valley in the Olduvai Gorge in present day Tanzania. These were dated to a period over a million years later than Lucy. They were given the name Homo habilis—handy man—because of the discovery of primitive stone tools in the same geologic layers. Habilis was taller than Lucy, a bit more human in skull and teeth, and with a brain case larger than Lucy's but still smaller than ours. Success in the struggle to evolve a large-brained individual was showing up in the fossil evidence. These could have been a later group of émigrés, or they could have been descendants of Lucy's kind. In a million years, we could expect significant changes in either scenario. Indeed, there are other fossils of robust hominids that have been found much farther south near the Cape but dated to a similar time period as habilis. These indicate that there were a range of different groups who settled and evolved in a range of different destinations.

The next group of fossils that are sufficiently different again to be given a name as a new species have been dated to more than a million years ago. This group was taller again, almost as tall as us. Brain size was larger again but still intermediate between Homo habilis and modern humans. This group was dubbed Homo erectus—upright man. The changes in the bodies of successive species reflected the evolutionary pressures that our ancestors experienced. Altered dentition and upright walking appeared earliest. Increased height took longer. And brain growth took the longest of all.

Then, about six hundred thousand years ago, a hominid suddenly appears in the fossil record who had a cranium as large or even a bit larger than ours. This creature became a very successful colonizer with similar fossils found not just in Africa, but as far away as Java (Indonesia), Beijing, Europe and the Middle East. The changes in anatomy that allowed larger brains had taken place. This led to a population explosion that propelled these people around the world. They colonized down the river valley as had their predecessors, but they also successfully headed north, either along the coast of the Red Sea or across the Sahara which was in a wet period, thence crossing into Asia, following the coast both to the Far East and north around the Mediterranean to the Middle East and Europe.

This species looked much like us. They were tall, probably hairless if we accept an aquatic phase in their past, with a cranium as large or slightly larger than that of a modern person. But there were still great differences. Jared Diamond described a Neanderthal, one of this group of peoples, thusly:

2. The Originator

Their eyebrows rested on prominently bulging bony ridges, and their noses and jaws and teeth protruded far forward. Their eyes lay in deep sockets, sunk behind the protruding nose and brow ridges. Their foreheads were low and sloping, unlike our high modern vertical foreheads, and their lower jaws sloped back without chins. Despite these startlingly primitive features, Neanderthal's brain size was nearly 10 percent greater than ours!

… While a Neanderthal in a business suit or dress would attract attention today, one in shorts or a bikini would draw gasps. Neanderthals were more heavily muscled, especially in their shoulders and neck, than all but the most avid bodybuilders. Their limb bones … had to be considerably thicker than ours. … Even their hands were much more powerful than ours; a Neanderthal's handshake would have been literally bone-crushing. While their average height was only around five feet four inches, their weight would have been at least twenty pounds more than that of a modern person.

… Neanderthal tools may have been simple hand-held stones not mounted on separate parts such as handles. The tools don't fall into distinct types with unique functions. There were no standardized bone tools, no bows and arrows.[1]

They were quite a bit like us but still very different. To live safely back on land as the upright hairless ape, evolutionary pressures had selected for individuals who could rely again on strength, not just on tool use.

And then, a little more than one hundred thousand years ago, another hominid species burst onto the scene. This creature was finer-boned and much less heavily muscled than the Neanderthals. This was us, a modern human, Homo sapiens. Fossils from this group exhibit changes to the shape of the skull that produced both a forehead and a chin, finally making them indistinguishable from you or me. And along with this new look came an explosion of tool artifacts. Where previously our Neanderthal relatives had shaped rocks to hit at something and could probably wield a wooden spear or club in their defence, we now find carved bone and antler tools fashioned for very specific functions such as fish hooks or needles, or flaked points shaped to be attached to wooden shafts. Within sixty or seventy thousand years from the first evidence of their existence, these people were producing the paints and the designs that grace the caves at Lascaux. The change is so dramatic that Jared Diamond called it, with little exaggeration, the great leap forward.

Another momentous change had been forced on the species. A particular group must have been trapped in an area from which they could not migrate and then stressed again by a changing habitat. Those few individuals that managed to pass through this genetic bottleneck did so by harnessing the inherent capacity of their brains, that organ which had been evolving for millions of years, growing ever larger with ever more complex cells and interconnections. The crisis forced a set of changes that appear in the fossil record as a forehead and a chin, a more gracile body and a huge increase in tool use.

These people overcame threats by thinking, by understanding their situation and fashioning responses. They adapted strategies instead of bodies to fit themselves to a variety of environments. When they finally met up again with their heavier stronger cousins, they could relate to their environment in an entirely different manner. They were proactively creating niches rather than looking to find one. Their arsenal of tools soon came to include fish hooks, needles, flaked points, bows and arrows, cave paintings, jewellery and clothing. Any or all could be called on depending on the situation at hand. Being able to solve problems was a spectacularly successful stance. They thrived everywhere. They migrated into and tamed almost every habitat on earth. The Neanderthals were sometimes absorbed and sometimes displaced, but they disappeared as an identifiable species in the face of this new competitor.

The change in the forehead shapes a space that houses the frontal cortex, that part of the brain that controls both reasoning and complex behaviours. New abilities sprang from this entirely new region of the brain. Also, individual brain cells of people today are quite different from those of the other primates. As the anthropologist David Pilbeam wrote:

> The changes in the brain of higher primates are found mainly in the cerebrum or higher centres. This part of the brain has little to do with basic running of the body, but rather with the coping with the environment... As well as structural changes in our brains, human brain cells are also different from those of apes; they are larger, more complex, with more interconnections.[2]

If the changes in cell complexity had already occurred by a few hundred thousand years ago, then the great leap could simply have been a case of a creature finally mastering a tool that it already possessed. It is also possible that the nature of the leap involved the development of a more sophisticated type of brain cell with more interconnections.

Another interesting possibility, consistent with other observed changes such as the more gracile body, the narrower face and the more effective tribal structure, is that the essence of the change may have been a further slowing of neural crest cell migration. Having a stronger reliance on our fellows forced on us could have taken the taming of ourselves to a whole new level. This would have had the effect of increasing our impulse control and further reducing the stress of living in groups, allowing us to act in much larger units with the plethora of benefits that would flow from more effective cooperation and specialization.

Choosing between these mechanisms is not necessary for our story. It is evident that this leap occurred. It is evident what the changes entailed. By a hundred thousand years ago, modern humans, with our new brains and our gracile bodies, our tools and our impulses to create art, had arrived. The slow movement to more cooperative, more thinking creatures that had been creeping along for millions of years had reached the point where a sudden leap could occur.

The shallow sea on the Danakil flats dried into a salt flat a few hundred thousand years ago. The increasing concentration of salt as the lake evaporated would have caused the lives of anyone remaining to become more and more untenable. Probably, the stronger and more adventurous left early, all of these being gone by six or seven hundred thousand years ago. The weaker, the more delicate, the more fearful were left behind to die. This, in itself, would create a genetic separation favouring the more gracile and least aggressive. Evolutionary leaps are made by the failures of the passing world, not by the victors.

When any of these hominids ventured onto the continent—Lucy's troop about three million years ago, Homo habilis about two million years ago, Homo erectus about a million years ago, Neanderthals about half a million years ago, or modern humans a little more than a hundred thousand years ago—they found themselves forced into a hostile world for which they were no longer suited. Their time on the island had changed them. They now faced large predators as creatures who were larger, slower, weaker and less fitted for climbing. They needed to find new food sources to replace their seafood diet. But they also had new strengths. They were used to acting as a group coordinated by shared rules and taboos. They possessed a facility with language and an ability to construct and use tools and weapons that went beyond that of any of their ancestors. Abilities in each of these areas were greater with each successive wave, but these advantages were common to them all.

They had to rely on the group to repel predators. This demanded that every member maintain constant vigilance and awareness of the positions and dispositions of their fellows, a stance they had already accepted and honed for internal safety. They found that rudimentary spears and clubs, wielded by a dozen people in an organized communicating manner, could repel big cats. Food gathering in a hostile environment forced them to move and act as one. The ability to cooperate had been forced on them in their search for internal safety, but this now became key to carving out a place amongst stronger, faster, fiercer predators.

Neanderthals were a large, strong social creature who could use language and tools. They had harnessed fire as shown by evidence of hearths from about five hundred thousand years ago. This would require cooperation by many people—some to gather firewood, some to protect the hearth, others to continuously tend the flame. Fire was a product of a successful group. They fashioned rudimentary clothing that could be created without needles. This would require hunters to capture animals and a protected base where others could prepare the pelts. Fire helped them survive extremes of weather, but it also allowed extraction of more readily assimilable energy from food which, in turn, would facilitate further growth, especially in the nurturing of brain cells. Fire also encouraged the sociability of the hearth further spurring the process of self-taming.[3]

Darkness was their most vulnerable time. Their vision was limited at night, while many of their predators were nocturnal. A defensible redoubt into which they could retreat from sunset to sunrise was essential. Caves worked well, allowing a few lookouts at the entrance to protect the rest as they slept. Our ancestors became cavemen by necessity. From caves, Homo sapiens advanced to barrows—constructed cave-like semi-underground dwellings approachable only through a narrow opening. This allowed settlement to follow where water and food were plentiful. Only much later, when they had become confident that their weapons conferred adequate individual defensive abilities, could they rely on freestanding dwellings such as tents or huts.

Each troop of Neanderthals created a survival strategy involving a defensible home base, procedures that organized defence, and skills and tools that provided adequate food. Traditions and taboos then codified these proven methods. These taboos included a discouragement of further change because, once stability had been achieved, further change would be a threat. Elaboration stopped. Because they worked well enough, Neanderthal tools stayed the same for hundreds of thousands of years. Their survival strategy was successful but static.

For humans after the great leap forward, such taboos on innovation were impossible to accept. Using novelty to surmount obstacles was the heart of their survival strategy. They could reflect on their behaviour, their successes and their shortcomings—a glance at the paintings on the cave walls illustrates this capacity to contemplate both their lives and their environment. Their great strength was that they had become planners. Which tool or technique would be appropriate here? What would my mate understand if I spoke this word with this construction? The possibility of choice in all aspects of their lives stimulated their curiosity and varied their activities. All animals can be curious about specific things, but Homo sapiens took that to a whole new level. The new brain created a curious animal.

Their survival strategy relied on versatility instead of on a static set of rules. For example, a conflict with Neanderthals would see the Neanderthals gather facing outward with their clubs as they had always done in facing danger, as their customs demanded, while this new species would be conferring, making choices about a range of weapons and tactics—bows and slings and spear-points, feints, ambushes and sieges—adapting and devising specific tools and strategies that were relevant to this specific confrontation. The thinkers thrived. Thoughtful choice turned out to be much more powerful than a skillfully repeated pattern. The traditions defining behaviour had to be less rigid for them. Only recognized dangers had to be prohibited and anything not specifically prohibited was expected to become the subject of variation. There was a real danger here. Innovation and change had become essential to their success while survival strategies had to be rigid enough to preclude any possible unforeseen consequences that could arise from innovative behaviours.

When a tribe's time was fully dedicated to providing food, shelter and safety, the urge to be curious was unlikely to lead very far. Repeating safe choices would lead to food while trying something new risked leaving you hungry. But in times or environments when food was plentiful, these people were fed and safe and bored. The urge to be curious led them to discoveries. If it was obvious that these innovations carried a risk to stability, they were proscribed. But there was a steady growth in what they knew, in how they did things, in the range and subtlety of their language and in the sophistication of their tools. Curiosity, and the new brain that had spawned it, was continually pushing their societies towards greater complexity.

These people could live successfully almost anywhere. That allowed their population to grow, forcing them to continually split and send subgroups to migrate beyond their borders. New colonists would always start with old strategies but, when these proved imperfectly matched to new conditions, their reservoir of knowledge would yield ideas that became important parts of their new lives. This flexibility could also be important if their environment changed. Knowledge of a new area, a new food, a new tool, a new weapon, or a new method of hunting, could spell the difference between famine and survival. Having the fruits of curiosity from the past did not guarantee survival, but no knowledge beyond their basic survival strategy meant no margins and a lack of margins was always risky.

Imagine that you are a member of such a group. Your tribe previously relied on hunting parties to deliver large game. They survived a collapse of the herd by turning to a hardy plant whose edibility and methods of preparation were known because, somewhere in the past, someone had felt driven to explore new tastes—a prehistoric Julia Child. These plants had previously been ignored but they now kept the tribe alive until a wider range of foods, perhaps small animals captured with newly invented snares, was available. Without this superfluous knowledge, the tribe would have perished. The folk memory of the group emphasized that this was not the first time that this had occurred. The cycle of crisis, adaptation, survival and the recrystallization of a new survival strategy had been part of their story over thousands of generations. Knowledge, uncovered because someone had been curious, had turned out to be essential more than once. An appreciation for the exercise of curiosity grew stronger over time and the traditions, rules and taboos had to adapt to reflect this. In many aspects of their lives, experimentation became actually encouraged.

Of course, some tribes did disappear in each crisis. There were never guarantees. But the tribes that contained the greater number of curious members, the tribes whose traditions had come to be more accepting and encouraging of experimental play, and the tribes whose survival strategy and bountiful environment had allowed the leisure to be bored, always had more options. They were more likely to survive. This process, repeated over and over again, had three related effects.

First, there was a selection for individuals who had an instinct to be curious. Even today, there are some people who have inherited a nature that seems predisposed to ask *why?* while others seem content to happily accept whatever they encounter. Tribes where most individuals could function well only when repeating a learned and accepted pattern would be the most likely to perish whenever conditions suddenly changed. Tribes who had the greatest number of members who had a more active sense of curiosity were more likely to have created the margins which would allow survival. The instinct to be curious grew stronger in the species as a whole.

Second, each major failure of the survival strategy ushered in a short period when experimentation was driven by desperation and unreservedly embraced. The old ways had proven inadequate so there was no longer any survival strategy to preserve. When you are starving, taboos against eating certain foods no longer have strength. In a few generations, when success had been found, a new survival strategy would evolve and unrestrained experimentation would again be suppressed, but while the crisis roiled their world, there were no limits. In these flurries of unfettered activity, many quite radically different survival strategies evolved, welding new discoveries and unused knowledge to valuable aspects of the past. Human survival strategies became ever more varied and more elaborate. Tool use, hunting methods, accepted foods, language and social organization became ever more complex.

And thirdly, an appreciation for experimentation, for curiosity and for individual discoveries were gradually incorporated into the traditions of the tribe. As survival became assured time after time because of novelty, it became impossible to label innovation as harmful to the group unless a specific reason could be advanced. The traditions grew to provide not only outlets but esteem for the fruits of individual curiosity. Individuals needed creative outlets and the traditions of the tribe had to adapt to take this into account.

As the urge to be curious strengthened, these people became Originators. Instead of embracing change only in the throes of a crisis, they came to court new horizons as an integral part of their everyday lives. Avenues were created by which the results of experimentation could make their way into tribal life. And as we spent our time on those pursuits, we spurred further changes in the brain. Usage shapes form. The individual who lifts weights changes the form of his body; the individual who uses his brain more rigorously encourages pathways between brain cells and a greater capacity to think.

The urge to originate expressed itself in many ways that could be readily valued within the tribe. Some people were curious as to what existed over the hills on the horizon. This knowledge had been useful in the past and so the tribe valued such knowledge for its own sake. If someone discovered a previously unknown river or sea, a place with new types of nuts or berries or fish or game, that knowledge could improve their lives. If it was a river that flowed to warmer or cooler regions, then that corridor could, in the future, become a pathway to survival. We became explorers. The great explorer earned the esteem of his fellows.

Some were curious to know the world of plants and animals more intimately. This knowledge was also very useful to the tribe. Every plant offered possibilities as food or medicine. The more they knew about animals, the more successful hunters they would be and the safer they would be from predators. Traditions evolved to value such knowledge. The person who knew about plants would become a healer or a preparer of foods. The expert on animals would lead the hunt or organize defence. These were also ways to earn the esteem of the group.

Some were curious to experiment with tools or weapons. If new tools could catch game that had been uncatchable, that would be a boon. Better weapons would help make all hunts more successful. Tools that allowed the tribe to use more of the prey, such as the sinews, bones or hide that had previously been discarded would give them more options and more ease. The expert in the production and use of such tools gained esteem. Traditions encouraged innovation in these areas. Taboos now only squelched that experimentation which could be clearly seen to threaten the core of the survival strategy.

Curiosity prompted tribal members to reflect on the world around them and on the nature of their existence. They saw and experienced birth and death. They experienced the cycles of day and night and the seasons of a year, this flow inspiring questions of how life began and how it would end. They noted their similarities with the other animals but how much they still seemed to be different. They created rituals, myths, songs and stories to ground their lives in this world. Some people had skills in memory, dramatic abilities, or susceptibility to entering trance and they became bards and shamans, the keepers of the stories and the leaders of the rituals, another route to earning the esteem of one's fellows.

In times of crisis, starvation shattered the traditions. All taboos became toothless. Even the fundamental taboo on internal violence might be sloughed off. Starving people would feel the urgings of the Dominator to take what they needed from the weaker and to hoard what they had rather than to face the crisis as a group. They would be dissuaded from this only by a recognition of their stronger need for the group in assuring their protection from predators and internal violence. When the crisis was over and new food sources were once again secure, the tribe always had to reaffirm the basic taboos that embodied their coalition against domination and hoarding.

Taboos do not rid individuals of urges. They can only induce people not to act on those impulses. Successful strategies always redirected the urge to dominate into a struggle to garner esteem in ways that would be beneficial to the tribe. A hunter striving to provide more meat than others, a defender killing more enemies than others, an explorer seen to be best at expanding horizons for everyone, or a tool user more skillful than all others in the tribe, all felt the satisfactions of competition that soothed the Dominator. People could still be recognized as being better than their fellows, but in ways that did not have to leave someone battered and bleeding.

2. The Originator

We can now see the full palette of human urges, each one conditioned by specific experiences in our evolutionary history, each still capable of driving us to act in the present. We humans will choose which urge to follow depending on the traditions, taboos, habits and infrastructure we create to direct our group activity. Vast differences in observed behaviour from different cultures can exist only because activity can erupt from one or another of these drives depending on the particulars of the context we have created.

First, we need food and water. If people are starving or dying from thirst, all other considerations are moot. In extremis, everything will be focused on our immediate needs. If, however, an individual believes that these needs will be best met by contributing her part to the survival strategy of the group, then these fundamental needs just add power to group efforts. We can be avid farmers or fishers or specialists of any kind if we can believe that these efforts will lead reliably to food and water.

There are also sexual urges whose purpose is to lead us forcefully towards leaving progeny. This is another fundamental part of our animal inheritance. Taboos may be able to direct this urge, but they cannot deny it. One of the important functions of a social structure is to define acceptable non-competitive routes for this urge to be expressed. If a social structure fails to meet this challenge for most of society, then the community will surely fail. If, however, it connects people into families of some sort, then this drive will also just add strength and cohesion to the survival strategy.

We have a powerful urge that drives us to be a Dominator and a Hoarder. Our ancestors spent millennia in dominance hierarchies, groups where preferential access to food and mating partners depended on cowing others in combat, real or mock. That instinct still whispers to us its immediate and bloody solution to any troubles. We must never underestimate the power of this drive. We must never assume that we have outgrown it, that humans are nothing but sweetness and light who sometimes fall into bad company. All successful social structures, both tribal and modern, have invested a great deal of energy into taboos to deny these urges an outlet and to create strategies to deal with situations where this drive erupts into violence.

Another powerful urge impels us to be the Originator—creating, exploring or experimenting. We are the only species that courts new horizons even while we enjoy assured success. This urge is responsible for the glories of civilization, but it has also forced us to abandon comfortable niches in favour of risky ventures and potential disasters. This part of our nature can be seen as perhaps 90 percent blessing and 10 percent curse.

We are also driven to be the Cooperator.

When we jettisoned the dominance hierarchy and tamed ourselves, we strengthened an urge that pushes us to embed ourselves in an enveloping group. Cooperation in large groups became natural to us. We gained the ability to be friendlier to others and accept them as an essential part of our lives. For millions of years, we have lived in a constant awareness of our fellows, relying on and assured by their presence, ready at all times to leap to their aid.

As Cooperators, we need to exist within a group for our definition as well as our survival. The lack of a community causes us unease. To feel ourselves safely embedded within one promotes contentment. Just as the urge to dominate has given rise to a small internal voice urging us to attack, grab and run, this urge to cooperate has given rise to an internal voice speaking in favour of community that we have come to designate as conscience.

Lastly, like all species, we have an instinctive urge to protect our survival strategy by shunning change and clinging to solutions that have worked in that past. We are by nature conservative.

These urges—to breed and leave progeny, to dominate and hoard, to originate, to embed ourselves in a cooperating group, to protect our survival strategy by shunning change—are the wellsprings of all human actions. They arose from our evolutionary history and they are real. They cannot be denied or altered for the convenience of modern times. This is the nature of the clay with which we can mould our civilizations. Our varied societies take their character from people responding to these urges within the rules which we construct for ourselves.

These urges can often conflict with each other. The urge to dominate can challenge the urge to embed ourselves in community. The urge to safeguard the survival strategy can conflict with the urge to originate. We are complicated. Coordination of this complexity can only be provided by the beliefs, customs, habits, traditions and taboos that guide our behaviour. Through the last hundred thousand years, in every new territory or in every adaptation to new conditions, such a social system had to emerge and gain acceptance before a people could thrive. Throughout that period, societies became progressively more complex as the Originator continuously pushed against accepted limits, weakening one taboo after the other.

We developed progressively more sophisticated weapons, which allowed individuals to defend themselves more effectively. This gradually eased the requirement that the group be constantly vigilant, ready to spring immediately to the defence of any threatened member. By about ten thousand years ago, individuals could count on being able to defend themselves against most threats without mobilizing the group. In many other areas, sophisticated tools also allowed individual activities to replace communal efforts. Individuality was creeping into tribal life. The feudal era was about to begin.

Chapter Three

The Originator Breaks Free

⌁➤

As the urge to be curious became a more significant human trait, tribes were gradually transformed. A few hundred thousand years ago, these people would have possessed a few replaceable tools made from sticks and shaped rocks. They would have had rough hides to wear, no shelter but what they could find, no foods preserved against hard times. These meagre possessions could be carried when the tribe moved or they could just as easily be abandoned and recreated later. But as tools became more varied and complex, more time had to be invested in their production and they became treasured. People were reluctant to abandon hides they had cured and decorated and sewn to become clothing or shelter, and foods they had carefully preserved to be stored against dearth. These were so obviously useful that the tribe's traditions had to change to give value to possessions.

In order to give value to these possessions, however, the tribe's traditions had to value curiosity itself. And this was where it got tricky. Seeking esteem had to be encouraged in specific areas where the benefits could be communal such as knife making, hide curing, clothing production, food preserving, tent making, warrior success and storytelling while still denying it wherever it threatened to lead to hierarchy such as in the personal amassing of goods. In time, people who had won esteem became stewards of the knife, experts in the curing of hides or foods, healers, leaders of a war party or shamans, encouraged to both excel and gift the benefits to the tribe. Since the urge to innovate could no longer be denied, it had to be controlled in ways that both provided benefits and preempted conflict.

But these tribes had to live in a world of droughts, floods, earthquakes, volcanoes, meteor impacts, ice ages and the unintended consequences of their own activity. Each time a crisis made their traditions irrelevant, as periodically did happen, they had more options to call on as they put the tribal structure back together. Every circumscribed possibility, every known skill, every arcane bit of knowledge, could then become an important part of their new world. Past innovations, no matter how much they had been denied or accepted, created new options. In these periods of trouble, people looked to the stewards and their skills. A more dominant position within the tribe briefly opened up to them and their latent drive to dominate would urge them to seize it. As security once again became assured and as new traditions emerged, the strengthened position of the stewards generated pressure to have more of the benefits of their skills and efforts stay with them. Stewardship slipped towards ownership. For common objects like clothing or tents that everyone could produce, greater differences between individuals became the norm. But to be safe in a dangerous world they all had to be able to stand together. The new traditions always had to curb individuality enough to reassert this cooperative stance.

However, as sophisticated knives, spears, arrows and bows became widespread as personal possessions, individuals were progressively better able to protect themselves from predators without needing the constant presence of the group. People no longer needed to be constantly with their fellows to be safe. The traditions demanding the primacy of the group could start to relax.

Each time a set of new traditions was put together after a crisis, the sphere of individual endeavour was larger than before, the tasks that must be approached as a group correspondingly less. The tribe needed people to be experimenting, exploring, making better tools, dreaming stories, and creating useful objects, so these needs drove the direction of change more than the fear of predators.

Groups came to differ drastically from each other, each with their particular traditions and their particular basket of skills adapted to their own particular environment. Success allowed a tribe to grow and when numbers made the group unwieldy, a subgroup would venture off into new territory. Modern humans spread over the globe. Adjacent tribes still retained connections, meeting regularly where their territories overlapped. At these meetings, mates could be found and trade, socializing and storytelling would take place. Tribes had neighbours.

However, it was also possible for a group to meet a tribe to whom there was no remembered connection. Taboos prohibiting violence did not apply to these strangers. There were no shackles on the Dominator in this situation. Violence directed at strangers was actually encouraged: warrior success was a sure route to esteem. Some practical goal like the protection of hunting grounds may have provided an excuse for hostilities but, because people possessed the urge to dominate, they found it natural to want to wage war. Since the whole tribe shared in any spoils—new hunting grounds, captured slaves or stolen tools and food—traditions came to grant significant esteem to a warrior, a fighter, a killer, a Dominator.

By about fifty thousand years ago, these people were settled in western Asia, Europe, the Middle East and Africa. By forty thousand years ago, all of Asia as well as Australia had been colonized. By twenty thousand years ago, they had reached the Americas. Each region they entered held new challenges and so conditioned a variation on the basic palette of skills. Humans proved to be very adaptable.

Evidence about the lives of peoples after this point is much more plentiful than it was for our more distant ancestors. Tribes were more numerous so there are more sites of habitation to study. Their relative newness also makes these sites easier to find. They are likely to include artefacts, middens, hearths, building remains, burial sites and art works, all of which describe important aspects of a life which can only be more vaguely inferred from fossils.

There is another source of evidence to help us visualize the lives of these people. A few hunter-gatherer tribes were sufficiently isolated to maintain their traditional patterns until fairly recent contact with modern societies. We have gained some knowledge of the pre-contact lives of the Bushmen or !Kung in the Kalahari, the aboriginal peoples from Australia, and a variety of tribes in the Americas, amongst others. There is a continuity between the lives of these tribes and that of earlier stone age peoples. Though tens of thousands of years of elaboration have introduced changes, the core of the survival strategies will still be consistent with what they once were. Common elements shared by these dispersed groups also point us to features of lives that must have existed in the past. At the very least, any conclusions we posit about the lives of our distant

Stone Age ancestors cannot be trusted if they contradict the lives of these survivors.

An important feature common to all these tribes is the importance of the integrity of the group to every aspect of the lives of its members. Methods of food procurement and distribution, assignment of mates, organization of defence, decision-making processes, religious beliefs and ritual practices promote the unity of the tribe at every point. Much care is taken to eliminate opportunities for disputes. Tribal identity is consciously reinforced in rituals, shared tasks, taboos and the design of living spaces. It is inconceivable that tribal unity did not have similar importance a few tens of thousands of years ago when less sophisticated weapons made cooperation even more vital. The brutal Hobbesian war of each against all does not describe life within a tribe. In fact, just the opposite is true: great care is always taken to assure a spot for all in the bosom of the group. Our contention that early humans were defined by their ability to maintain a cohesive group is bolstered by every example.

The myths, rituals, customs, taboos, kin relationships, language, art, tools and technologies created a world which enveloped each person. An individual life had no meaning apart from the life of the community. Conversely, the community was constantly being reinforced as a living entity by every action of each individual. There was no division between community and individual: they were two sides of the same coin. The modern concept of an identity crisis had no meaning in this world. This was the essence of every human life for tens of thousands of years, for thousands of generations.

The great provider is a person of status in every tribal society. Someone who can produce much, or who can persuade others to work with him to produce much, gains esteem which can mature into prestige. As an example, among the Plains Indians of North America, Robert Lowie wrote:

> … on the plains, … a great man could maintain his status best by lavish generosity to the poor. Such liberality, next to a fine war record, was the basis for high standing. The Oglala had a society of chiefs enjoying superior prestige, but when a novice was admitted, he was urged to look after the poor, especially the widows and orphans. Among the Blackfoot, a man aspiring to become a leader tried to outshine his competitors by his feasts and presents even at the cost of impoverishment. … As a rule, the chiefs were titular and any power exercised within the tribe was exercised by the total body of responsible men who had qualified for social eminence by their war record and generosity.[1]

Since every male was a warrior, however, no one could force others to help him amass the goods to gain his prestige even if such an idea could have occurred to him. Warriors could sometimes be led but could never be driven.

Quite a large area was needed to assure that a tribe of hunter-gatherers had access to adequate food in all seasons, good years and bad. Populations, though spreading over the globe, were never large by modern standards. The anthropologist Marvin Harris estimates that "… in all of France during the late stone age, there were probably no more than 20,000 and possibly as few as 1,600 human beings."[2]

If each tribal unit averaged twenty to fifty members, there would be somewhere between 100 and 1,000 separate tribes in Stone Age France. Since this population density was enforced by the amount of food available in periods of scarcity and since these times of scarcity probably occurred only in a specific season in an unusual year, the normal state for these peoples was one of abundance. As studies have shown about the !Kung:

> ... despite their habitat—the edge of the Kalahari, a desert region whose lushness is hardly comparable to that of France during the Palaeolithic period—less than three hours per day per adult is all that is needed for the Bushmen to obtain a diet rich in proteins and other essential nutrients.[3]

Stone Age people, for most of their year, had time that was not directed by their need to find food. They had time to be curious. This guaranteed a steady stream of new innovations. The steady movement towards individuality was powered along its inevitable route.

Two related discoveries that subsequently became important for us were cities and agriculture. Journalist Jane Jacobs described a series of small innovations that would permit hunter gatherers to develop both.[4] Her argument that cities preceded agriculture is based on her reading of the archeological excavations at Çatalhöyük, a city that existed on the Anatolian plain by nine thousand years ago but whose origins are older than that. None of these steps are improbable. In fact, for the Originator, they should be expected. Similar opportunities led independently to similar results in Southern Iraq, in Southeast Asia, in Southwest Asia, in China, in Central Mexico, in South America, and in Sub-Sahara Africa. The process was as follows:

Obsidian, a rare volcanic glasslike rock, exists in this area of Anatolia. It was much prized because it could be worked into points and blades superior to anything else then available. The tribe whose territory contained this resource would inevitably discover it on their travels and, over time, would experiment in methods of extraction, fashioning and use. They would enhance their own lives with improved tools and weapons. Adjacent tribes, noting this at their gatherings, would want to trade for either pieces of obsidian or for the tools themselves. Because obsidian was rare, valuable items would be offered in exchange. News of this substance would spread from tribe to tribe and more distant groups would want to trade with intermediaries or attempt to mount trading expeditions to deal with the miners themselves. Negotiating permissions to cross intervening territories would provide benefit to these tribes as well. The surprise would be if these actions had not occurred once such a rare item had been discovered.

An outcrop can easily produce much more obsidian than one tribe can use so producing the excess for trade would not be difficult. Trading would provide them with goods that either did not exist on their territory or that they would have had to work harder to produce than they did in gathering the obsidian. Mining and trading would be added to their traditional tasks of hunting and gathering. The appearance of trading partners would also mean that not all strangers were automatically seen as enemies—some strangers could bring benefits to the tribe. Other useful but rare items would have initiated similar scenarios. For example, tribes on the coast could trade seafood or shells with neighbours in the interior. Tribes with deposits of salt would have had similar opportunities.

Each successful trading expedition would spread the knowledge of obsidian further. Trade routes of ever-increasing length evolved. Along these routes, intervening tribes sold provisions and access rights to travellers. We have evidence of such Stone Age trade routes.[5]

Temporary settlements provided space for the actual trading to occur. This would be distant from the mine site so that the source of wealth could be kept secret. Strangers would camp on one side of an open area while the local tribe would be on the other, meeting to effect the exchange in a central space made safe by agreed customs, taboos and rituals. It would take effort to produce a set of rules that could overcome the natural hostility towards "the other," but it could be done.

Those seeking obsidian, being hunters and gatherers, would bring the surplus of their own success to trade—meat, animals, skins, fruits, seeds, nuts and other plant products. Since meat spoils easily, it would be logical to capture the most docile wild animals, the forerunners of sheep and goats and fowl, and bring them in cages, on tethers or in herds to offer live meat as part of the exchange. The locals could use these animals to replace their own efforts in hunting.

Because travel is easier at a particular time of year and because it is only possible to gather surplus of many products in the appropriate season, trading was initially seasonal and the market square was a temporary encampment. Many trading sites evolved to this point and no further. However, if the product was in great demand, as was the case for obsidian, the local tribe received many animals during the trading season. A few members of the tribe kept these animals contained and healthy and protected from predators. This necessitated upgrading the temporary enclosures into more substantial holding areas. The rest of the tribe spent more and more time close to this food source. The size of the encampment and the length of occupation increased. The animals' offspring further augmented the food supply. Some females were kept from slaughter to the very last in hopes of this extra gain.

If animals could be kept through the season where hunting was traditionally most difficult, the tribe could grow beyond the limits imposed by the richness of their territory. As tribal numbers grew, they soon needed more food than hunting could provide, so their reliance on trading would increase. Hunting would become less important because they had the traded meat. More people and less hunting meant everyone had time to contribute as miners, traders, granary overseers or animal keepers. The temporary trading encampment would evolve into a permanent settlement. Husbanding of animals would start to occur independently of the trading that took place.

At the same time, people brought grains and seeds to the trading area from many different regions. These grains had to be stored, necessitating the construction of granaries. Around the granaries, accidental scattering would cause a variety of grains to grow in greater density than would have occurred in the wild. This not only showed what could be available from densely-planted crops, but it promoted cross-fertilization resulting in new varieties, the best of which would be noted, saved and deliberately scattered.

The presence of a rare commodity would be all that was needed to lead a curious people to permanent settlements with divisions of labour such as animal husbandry and horticulture. The settlement would also need contributions from many other skilled individuals such as builders or tanners. As population increased, the tribe could intensify agriculture instead of sending a subgroup off as pioneers. As distant trading peoples visited the market square, they encountered these new ideas of farming and division of labour which they could incorporate in some measure into their own lives, either immediately or during some future crisis. Knowledge of horticulture, agriculture, trading and the crafts that supported them became known throughout the tribal world even as most tribes still protected the centrality of hunting and gathering.

When the glaciers retreated about ten thousand years ago, climate changed everywhere and most survival strategies were thrown into disarray. Tribes aware of horticulture and animal husbandry could call on these skills. In a few places, this provided the road to survival. But it came at a cost. In every crisis in the past, the integrity of the tribe, with everyone acting as one in defence, had been reaffirmed as the cornerstone of tribal organization. However, when agriculture replaced hunting as the main source of food, this type of defence became more difficult to provide.

Crucially, they could no longer retreat from danger. Farmers have to tend a crop all year long and fields cannot be moved. The tribe had to protect stored harvests until the next one was due because these were all that stood between the tribe and starvation. Herds required secure, defensible enclosures. The ability to defend the granaries, fields and herds was fundamental. Defence from wild animals was easy; defence against hostile tribes was more difficult. War became a very different proposition than it had been when each group could advance, retreat or migrate until new tribal boundaries that accurately reflected changing power relationships were established. Having land that could never be lost was the central concern for tribes when they reestablished new traditions based on farming.

Choosing fishing and farming as a strategy would not work everywhere. The best sites, usually river valleys with natural irrigation and flood plain fertility, became highly valued. Groups claiming such a territory could provide abundant food and grow while other tribes had to retreat into the hinterland to remain small and nomadic. The ability to defend a good site separated the winners from the losers.

Defence was complicated by another fact: men who gathered, fished, farmed, planted and weeded no longer practiced and honed the skills, strength, wildness and mindset of a warrior as they had during regular hunting expeditions. Men were no longer skilled in the use of weapons as a matter of course. The answer to this dilemma was found in the concept of division of labour that was an essential part of the farming strategy. A cadre of individuals were chosen to hone the skills of warring instead of the skills of farming, herding, trading, mining or fishing. If warrior skills were not emerging naturally in daily activity, then they would have to be instilled through training. A group of members were given the task of defending the tribe. This arrangement worked. The tribes that first hit on this scheme thrived and claimed the most fertile territories.

The nutritious, plentiful and year round food produced in a river valley allowed an increase in numbers far beyond what was possible for a nomadic hunting tribe. In an agricultural community, more people meant more builders of barns, more keepers of cattle, more fishermen and more field workers, all leading to more food. And this larger population also allowed more men to be designated as warriors.

Of course, the extent of the arable, fertile land still imposed a limit on the population. But now, if a tribe reached the carrying capacity of their land, they had a new option to consider as well as the traditional one of splitting and sending out colonists. Though the warriors had initially been charged with defending the granary, their skills could just as easily be directed outward to drive neighbouring tribes off their fields, seize stored crops and take over developed farms. Expanding their territory allowed the group to keep growing. The possibility of such an attack conditioned the development of every group in the area—the only way not to become a victim of your neighbour was to support a suitably strong group of defenders yourselves. It was the first arms race. The professional soldier strode onto the stage of history, strewing mayhem in his wake. He has held the spotlight ever since.

Tribal groups chose the most fearsome warriors as the first group of soldiers. They were aware of the old traditions mandating the sharing of goods and power amongst all in the tribe. They were aware that leadership meant impoverishing themselves to assure that everyone was fed, and they were aware of the traditions that harnessed every act to the well-being of the tribe. These traditions were strong memories, but the actual power relationships had changed.

The strength of the old traditions had rested on the fact that they reflected the actual distribution of power within the tribe. When every man was a hunter and a warrior, power had to be shared. Effective coalitions could readily form to demand it. Now that farmers and fishermen were weaker than soldiers, coalitions comprised of one group could not easily oppose members of the other. This inequality in strength and skill in arms ruled out a continuity with the old traditions. They became distant memories.

To defend the tribe, soldiers had to overcome the military might of any neighbouring tribe or village. This made them capable of subduing their own villagers. They were no longer one specialist amongst many—their particular specialization made them superior. Even within their own small group, force assumed an ever more prominent role with gradations in privilege emerging as alpha males once again jockeyed for primacy. The Dominator threw off his chains. The elites of the tribe created a dominance hierarchy.

If one warrior could feel the old traditions so strongly that he could resist the temptation to abuse his power, could he resist the desire to bring his son into his privileged position with him? And could his son, or indeed any soldier of the second, third, or subsequent generations, continue to honour traditions they had never experienced? How can you banish someone if the soldiers do not agree? New traditions had to arise that reflected the power distribution of the present, not the past. The central fact of these new traditions was that a small group of men could impose their will on the community.

There was an attempt to carry forward the injunctions that directed soldiers to use their strength for the good of the community. Myths about knights selflessly dedicating their lives to good deeds were stubbornly persistent, but these could never hold sway. The reality was that using force produced rewards. Aggrandizing myths and investiture ceremonies based on self-serving lies were just a veneer of public relations papering over the exercise of raw power.

The degree of dominance that soldiers exercised depended solely on how completely they controlled the food supply. In areas where hunting and fishing were still possible, villagers were not totally dependent on the granary. There were options should they incur the rulers' wrath. Widespread reliance on hunting also allowed some warrior skills to persist throughout society. The example of Robin Hood, the outlaw as a popular challenger to the soldiers in the castle, is typical of stories from many cultures that express this possibility.

However, in societies where all food depended on the harvest, control of the harvest provided absolute control over the people. Soldiers always tied the redistribution of food to obedience. They manufactured hunger by redistributing only enough to keep the villagers alive, using the rest to enhance their own lives by supporting builders, blacksmiths, armorers, weavers, potters, servants, jesters, musicians, whores or whatever. They reduced villagers' lives to constant work and food insecurity. And they transformed the vitality, independence and curiosity of the tribesman into the servile stoicism of the peasant.

In a tribal society, individuals always made their own choices between time spent providing for basic needs and time spent indulging their curiosity. Consumption was limited because the producer and the consumer were one and the same. The !Kung spent three hours a day gathering food and then they stopped to socialize, tell stories or follow other whims. However, when consumers are a different group than producers, increased consumption is no longer limited by a desire to use one's time differently. In a feudal society, the elite always saw a need for more food production because this meant more soldiers and more lackeys. The peasants were always pushed for more, no matter how full their days may already be. Consumption by the few replaced the curiosity of the many.

Much of the world, except for a few tribes that retreated into lands entirely unsuitable for farming, was brought into the feudal world either by creating an army to defend their own farmland or by being overrun by another village that had already taken that step. That ushered in a new phase in human history, one shaped by the appetites of the soldiers and the evolving interactions between the elites and the peasants. This is the context of the human story of the last ten thousand years. And this is the subject of the next chapter.

CHAPTER FOUR

The Growth of the State

The feudal era was born when a tribe in crisis chose a few warriors to work exclusively at being the defenders of the group. Once these soldiers were empowered, this deal could not be renounced. The inherent inequality between soldiers and non-soldiers transformed a tribe of equals into a small elite group of rulers and a subservient village of workers. From that point, the appetites of the elites drove the historical process. Stability was the first casualty. An immanent desire for more—more power, more wealth, more growth—became the hallmark of this new form of society.

The tribe turned into a village. A nomadic existence in a large hunting and gathering territory was replaced by specialized tasks which had to be performed within fields adjacent to permanent dwellings. The peasants still lived, worked, played, celebrated and suffered together. Many aspects of their lives still existed within this group of equals. The customs that had shaped their connections to each other in the tribal setting still made sense. Many were retained in some form: concern and generosity towards one's village, relatives and neighbours are salient features of all peasant cultures. The peasants could still express the urge to cooperate and to embed themselves in a nurturing community. The old taboos and the ability to form alliances to enforce them could keep domination in check amongst themselves.

The villagers relied on the soldiers to defend them from external enemies. But, being decidedly weaker, they also had to endure the arbitrary demands of this cadre of defenders. This was their dilemma. To gain protection from external enemies they lived at the mercy of a group that expressed the same propensity for domination as the enemies they were expected to repel.

The peasants had to produce the food that kept both soldiers and peasants alive, so it was in the interests of the nobility not to kill their peasants, and even to allow them to live well enough to be able to put in a full day's work and to raise the next generation of healthy workers. But even with self-interest staying the hand of the elites, there was much to be feared from an omnipresent bully. The peasants had no recourse if abused. They had only the bonds of fellow feeling within the community to provide solace after the fact. Random predations by the nobility joined the weather as a circumstance of their lives that was beyond their control, events that could neither be changed nor avoided, only endured.

Each villager possessed the drive to be the Originator. But instead of having leisure to indulge their curiosity, their schedule was at the mercy of the nobility. The harshness of the demands placed on them varied. In cases where peasants were plentiful relative to the work that needed to be done, feudalism could degenerate into slavery. Peasants were worked until they were worn out, then discarded and replaced by others. But when the total amount of food could be increased by having more workers, then healthy peasants were a valuable asset. If time given to peasants to till their own gardens and decorate their own homes, furniture and clothing made them more effective workers and more likely to raise more healthy workers, then these activities could become part of village life. Villagers could then strive for esteem as builders, skilled healers, singers or storytellers. In these circumstances, the Originator might have some scope.

The community as a whole was too weak to curb the excesses of the elite, so each soldier answered only to his inner Dominator, stealing what he could, up to the point where he was thwarted by more powerful fellows. The alpha amongst them emerged victorious from this free-for-all and they settled into a dominance hierarchy reminiscent of the lower primates before them. They had no way to produce a community amongst themselves except by force and they were decisively cut off from the rest of the village. They each stood alone. No longer could they participate in any religious ceremony except as a symbol of power. No longer could they be an artist or a storyteller or a shaman or a healer and expect esteem not to be coloured by envy and fear. Their drive for position in the hierarchy poisoned every activity that relied on equality. The peasants still had vestiges of this in the cohesion of village life. The soldiers, however, only had continual struggle to maintain their position against challengers. One defeat or period of weakness and they would be cast aside. Living by power is a harsh mistress. For a creature who had, for thousands of generations, strove to embed themselves in a nurturing community, this was a great loss.

Your fellow soldier was not a partner except in the face of enemies in the heat of a battle. On most days, he was a competitor trying to rise above you on the greasy pole of power. To limit the carnage, they developed ritualized gestures of fealty and submission. These public affirmations of the current hierarchy shaped periods of temporary peace. But there was no equality. There was no trust. There was no reliable safety. There was not yet a cohesive military caste in which to root cooperation. There were none of the satisfactions of being the Cooperator that could still be available to the peasants.

The chances of a soldier gaining satisfaction as the Originator were not much better. The satisfactions to be gained from being a tribal leader, a warrior, a shaman, a healer, an explorer or a creator, were inextricably bound up with the esteem of the group. This esteem was now stained by fear. The nobility could still hunt, explore or follow solitary pursuits. They could seek the private exhilarations that arose from satisfying one's curiosity, but it was robbed of its savour by the absence of an enveloping group who would validate the accomplishment with esteem.

With satisfactions from acting as the Originator and the Cooperator both closed off, the only way the soldiers could provide meaning to their lives was to indulge that part of themselves that urged them to be the Dominator. It was not a question of winners and losers, of good guys and bad guys, of good and evil. Both the peasantry and the nobility were responding to the urges that made them human in the ways demanded of them by this new situation. Millions of years earlier, our ancestors had ventured down an evolutionary path based on our becoming a large-brained, thinking species living in cooperative groups. Because of that, we became the Originator, the creator of new options. Our great strength was our ability to fit ourselves to new situations. This strategy allowed us to thrive, but it had always contained a risk. Every success encouraged more individuality in a context where group solidarity was essential. This continuously increased the possibility that individual competence would become so robust that traditions would no longer be able to maintain the integrity of the group. Feudalism was a logical outcome of our becoming the Originator. The challenge of what to do with the Dominator is an inevitable part of the human journey.

No matter what else was occurring, sufficient food had to be produced to feed both farmers and soldiers. The elites gathered this food into central granaries for storage, protection and redistribution. Doling out this food gave them the power of life and death. The great provider had always been a route to status but now that contributions were forced, this role took on a very different aspect. The community was still enjoined to offer thanks but, in order to feed the soldiers and the growing entourage of lackeys, less food was always given back than was gathered. The great provider became a great reducer.

The elites chose who got to live without having to produce food. It could be more servants to cook, clean, carry and serve. It could be more artisans making and decorating pots, buildings and clothing. It could be more metallurgists and blacksmiths to produce more tools and weapons. Or it could be more soldiers. Whatever the choices, the surplus was explicitly directed to goods that had already been visualized. The surplus did not stimulate human curiosity to meander into unexpected corners.

The elites could always visualize uses for more surplus. There was no possibility of the villagers satisfying their masters and then taking time to indulge their own curiosity. Could a ruler have one servant without fantasizing about how his life would be improved with two? Could he have two soldiers without wanting four? The village was doomed to always be pushed to feed the largest possible noble entourage.

Initially, many tribes made the leap to farming. The ideas were known and there were many desperate tribes.

Each new settlement then had neighbours up and down the river valley who possessed both granaries and fields. The soldiers knew that this wealth, if captured, would increase their power. There was a constant whisper from the Dominator urging them to attack these neighbours. Animals kill to eat. Tribal societies made war to defend hunting grounds. In feudal society, men waged war to experience the satisfactions of waging war, of dominating, of killing, of stealing. They launched campaigns when food was plentiful and their granaries were full. War was a response to the urge to dominate, the one urge that the ruling class possessed to define and express themselves.

I stress again that these urges are real. We are physically impelled to do some things: to create, to be curious, to cooperate, to embed ourselves in a nurturing group and to dominate. We experience real satisfactions when we comply with these immanent demands. Humans have evolved neural structures such as endorphins-producers and serotonin receptors that generate physical reactions in response to specific activities. When our ability to gain such satisfaction is closed off for some activities, we will displace our efforts into others as we continue our search for the effect. If our context allows only one urge to be expressed, as was the case for the nobility in these early feudal times, then this route to satisfaction will be seized with a desperate force.

The villagers developed a docility in the face of the obvious disparity in strength that frustrated the soldiers' needs to be bullies. They knew they were needed alive and healthy to produce food. So, if soldiers wanted bootlickers, the peasants would lick boots knowing that they would soon be allowed to return to their farming tasks and their families. The satisfactions gleaned from domination rely on actually forcing people to do things against their will. The peasants were very unsatisfying victims. This directed the soldiers' aggression towards the military in neighbouring villages who had a lot to lose and would surely resist. They never launched attacks with logical considerations of risk versus gain regardless of what the history books try to tell us. Far from it. Any possibility of success would justify a campaign. Even futile sorties could look attractive to frustrated bullies. Dominators had to dominate. A man's got to do what a man's got to do.

Any community that adopted farming without empowering soldiers soon found themselves under threat from belligerent neighbours. They had to empower soldiers or end up conquered. The feudal structure spread through the settled world like a virus. Settled areas became divided into demesnes, each ruled by a group of soldiers whose perception of the balance of power was skewed by their need to dominate. Wars were inevitable. Violence defined the world.

In cases where neighbours were evenly matched, wars would generally degenerate into sieges that would peter out after a few deaths, much waste, destruction of fields and animals, loss of harvests and starvation on both sides. However, some wars did lead to a winner. A few of the losing soldiers would die, providing the ultimate affirmation of dominance for the winners. A few peasants who failed to keep out of sight might be killed but soon the old rhythms of life would return for them because their role as food producers remained. A trusted deputy of the victorious alpha male would take over the conquered castle to rule in the name of his lord. Ever larger regions were consolidated in this fashion, each village answering to a local soldier who swore allegiance to a distant ruler.

As these regions grew, they provided an ever-increasing number of soldiers ready to act at the behest of nobles at the apex of the pyramid. This perception of strength prompted the leader to attack ever-stronger enemies. Villages standing alone were overrun. Wars between equals became larger, longer, more wasteful, more deadly and more destructive. However, limits still existed to the degree that this consolidation could continue. Natural boundaries, such as mountains, a sea, or reaching the edges of fertile land, would stop the expansion. In some cases, as for example in ancient Egypt, a ruler could gain control of a whole area that had secure geographical boundaries. It was rare, however, for conquerable regions to be so conveniently isolated from the rest of the inhabited world.

The ability to control a large territory was the more common limiting factor. The ruler always had to rely on retainers at the local level. These men were also ambitious. Subservience grated, especially when orders came from far away. As regions became larger, there were an increasing number of dispersed soldiers eager to assert their independence. Groupings of such rebellious vassals were a constant danger. Further expansion made holding the territory together more difficult. Cycles of wars of conquest interspersed with wars of revolution eventually defined the boundaries of most kingdoms.

The most ruthless, the most vicious, those most strongly lusting after power, became the rulers of the feudal world. Gaining, extending and keeping power selected for murderous skills and psychopathic attitudes as these were most likely to lead to success. In the tribal world, leaders had been individuals of generosity, self-effacement, honesty and bravery. Even though the nobles claimed that those virtues still defined them, this was a new world with new power arrangements that selected for and conditioned new types of men.

There was little security for the elites. One loss in battle would cost them their life. They could be quite sure that any

4. The Growth of the State

subordinate would happily slip a knife between their ribs should it speed their rise in the hierarchy. The allegiance of these subordinates was maintained by fear alone and this fear could dissipate with advancing age or ill-timed disease. The regular wars and revolutions squandered resources leaving famine never far from either village or castle. Life at the apex of a dominance hierarchy had no gentle end point for the king of the castle—it was just one damn battle after another until they were settled into an early grave.

Eventually, even the most stupid nobles had to recognize that their chances of winning and holding further territories had become very slim. Their surviving neighbours had strong enough defences to withstand attack. Wars against them would weaken both, inviting challenges from within their ranks. They reluctantly had to move to legitimize the status quo.

This necessitated making agreements with neighbours which recognized borders and reduced other sources of friction. Internally, coalitions of the lesser nobility used the threat of revolution to force the king to accept limits on the absolute power he claimed over their lives. These coalesced into a code of rights and obligations that made the relationship between the king and his liegemen more predictable. This drama echoed our past where coalitions of middling powers had successfully opposed the violence of the alpha male and established the taboos which enabled a cooperative group. This produced the first code of laws.

The Magna Carta embodied this step in English history but some such limitations on absolute power began to appear in every feudal society that consolidated into a kingdom. If the elites failed to take this step, they would continue to tear themselves apart with rebellious struggle. There was no enforcement of these codes except from coalitions that could punish transgressors and these coalitions were always fraught. Members would strive to become too powerful to be punished. But the fear of a return to an unregulated free-for-all generated grudging respect for the code. Rebellions became more risky and more difficult to organize. And between these periodic episodes of bloodletting, a measure of stability could settle onto the kingdom.

Stories had to be invented to explain how a small privileged aristocracy was able to rise to a position above the mass of the peasants. The actual truth that their status had been seized by violence and maintained by force was not useful because it might encourage others to test their strength. So, to discourage revolt and to define themselves to themselves, a story was put forth that claimed that a small group of exceptional individuals had been chosen by the gods to occupy the apex of society. Over and over again in diverse cultures, kingship, once it had been consolidated by force, was presented as the will of the gods.

The religions this idea spawned went on to claim that the gods had also assigned a role to every person in the kingdom from the lowest serf to the king himself. If you accepted that you had been put on earth to fill a specific role, then you could get on with your daily duties without ever trying to change your lot. If you were a peasant, your life existed at the whim of the nobility and you had to work hard, but that was the life that your god intended for you. If you were noble, you were a member of the warrior class who was expected to risk his life to defend the realm if it was ever under attack. Between times you were expected to live a life of privilege. Everyone had an assigned role. All promised loyalty to the king whose edicts were the voice of god. And if you were king, you could expect to be obeyed in everything save those few restrictions codified in the incipient code of laws.

To inculcate this new creed, the king had to wipe out the old religions whose beliefs and practices stressed equality and inclusion, a way of seeing the world that was antithetical to the new story of god-sanctioned elites. It took an active army to make a belief in the divine authority of the king widely accepted. But making this state religion a part of everyone's life was another essential step in creating a kingdom that would endure. Without both an initial set of laws and a widely accepted mythology, a kingdom would fail to thrive.

Acceptance of divine will as the animating force of the hierarchical structure did three things. First, it birthed an aristocratic class bound more by fellow feeling than by fear. This class, of which each member could feel themselves to be an integral and included part, allowed its members to seek esteem in fields other than war. They could be artists, healers, visionaries, hosts and builders, and gain genuine esteem from their fellows. They could experience the satisfactions of the Originator again, even as many outlets for the Dominator were being closed off.

Secondly, when everybody accepted that the present order had been established by the gods, everybody *did* actually plug away at fulfilling their role, even if their particular cog in the machine was uncomfortable. If it was your position to till the fields, you no longer had to be bullied into doing that. If you accepted that it was the position of the nobility to produce ostentatious displays of wealth, then you could regard those displays without jealousy. You could take pride in the aristocracy as an embodiment of the wealth of the culture. If both peasants and nobility believed that god had created this situation, then the aristocracy no longer had to bully and the peasantry no longer had to live in constant fear.

In those cases where feudalism had slipped towards slavery, it was not necessary for the enslaved to believe in divine will. The two classes would always exist in completely separate worlds ruled by violence instead of agreement. For the nobility in these cases, specified borders, laws regulating the aristocratic class, and a story that provided meaning for them could still allow a cohesive class to form. This could still open some satisfactions to the Originator, though the masters would still always have the temptation to derive their meaning and satisfaction from dominating the enslaved. Even in a slave-based society, a stable state could be created if the overwhelming power differential could be maintained.

Thirdly, when the resources and the lives previously wasted in war became available to society, everybody actually was better off. This retreat of famine provided powerful validation for the emerging order. It benefitted everyone to believe this new story if that belief did keep rebellions and wars at bay and larger harvests in the granaries. The truth of the story was irrelevant. The existence of something that could order society without violence was what ultimately mattered most.

Nobles still felt the urge to dominate and those who lusted after kingly power were seldom dissuaded from rebellion by a belief that the king's position had been ordained by the gods. There were always reasons to believe that the wishes of the gods had been misconstrued in the past. If the rebellion was successful, that in itself was proof that the gods had wanted a change of dynasty.

The first code of laws defined relations between noble and noble and between noble and king. The peasants were given no protection from violent acts by the nobility. They had only their usefulness as workers to stay his hand. "If a man slay his serf, his is the sin and his is the loss," states the Domesday Book describing England in the twelfth century. Protection under the law extended only to those able to create alliances that were strong enough to oppose those who wished to dominate them.

Even though everyone, from peasant to king, believed in the right of the nobility to rule, and even though this shared belief provided stability for which everyone was thankful, the feudal structure would not have lasted for a minute without the presence of overpowering military force in the hands of the nobility. The peasants were continually reminded that the nobility were stronger than them, that the two classes were so mismatched in military might that any rebellion would be instantly and decisively crushed. Any coalitions to that end had to be understood as obviously futile. God may have willed it, but the nobility had to make it so. This preponderance of strength depended on two things—the number of soldiers that a noble could muster and train, and the wealth he commanded to armour them so that a few men could subdue a larger group of less-effectively armed and trained individuals. As long as the land could continue to produce enough food, and as long as the nobility could control that surplus to feed, train and arm soldiers, breed horses and produce armour, and as long as the armies so equipped could crush large groups of peasants, then the feudal order need never change.

When fewer resources were being wasted in war, more surplus became available. The peasants' lives did not change appreciably, but the elites responded to this new wealth by increasing their consumption. More artisans were supported. Some castles, because of their location on trading routes, their access to raw materials, or the greater military and economic power of their lord, commanded more resources and these courts attracted and supported more artisans. Towns, a gathering of individuals that was very different in form and function from agricultural villages, grew up around these castles. The urban world took its few first steps in response to the surplus of the feudal order.

These townspeople, unlike their village brethren who were dispersed around the demesne to be close to their plots of land, were densely congregated. In time, some of these artisans were able to orient their activities away from serving one specific lord and towards the trade routes and market squares that were becoming an important part of these towns. They arrived at a position where they could live without being subservient to a particular noble. They had slipped out of the feudal structure with its rigidly defined place for everyone. They were paid in kind or in some form of scrip for their work leaving the rest of their lives free to be lived according to their own wishes. Eventually, some of them would be able to move from court to court or from town to town as work presented itself. Some of these artisans and merchants accumulated significant wealth. This gave them, too, the possibility of feeding soldiers and buying horses and armour.

The towns became a force, not just because of the wealth of their leading citizens, but because of the numbers of people gathered densely together. A mob in tight streets—a new type of coalition—negated the inherent advantage of horsed knights. As towns grew, these people began to sense their power and they became less willing to suffer arbitrary violence or predation at the hands of the nobility. Bullying was seen as injustice, no matter what the religion claimed. The nobility were faced with a choice: they could either crush the towns or they could alter the structure of privilege in ways that co-opted this new power. Crushing was the choice in some instances, but this had a cost because they would then lose many important things that the towns produced. Often, the nobility bit the bullet and recognized the power of the towns. New laws were added to the code that granted rights to a person who was neither a serf nor a noble—a freeman.

Every check on arbitrary power added to the legal canon has been won by force or the threat of force. Our systems of laws are the result of compromises reluctantly ceded to potentially dangerous groups to gain their support for a system that, at its core, is designed to allow a privileged group to dominate. The idea that an intrinsic sense of right and wrong is the basis of the legal system is window dressing that persists because it is useful to the powerful groups the law favours.

When a system of laws favours those who have the power to enforce it and it does not disadvantage any groups who have sufficient power to rise in protest, it can produce stability. This stability arises from the fact that the laws are an accurate reflection of the distribution of power in society, not from anything intrinsic to the laws themselves. When power relationships change, because of new weapons or demographic changes or some other force, then the laws must adapt so that the privileges and protections of the laws apply to the rising power. If this is not done, the newcomers will demand their rights and the privileged group will have to defend their position. If they fight and are defeated, they face losing their position entirely instead of only having to share.

A system of laws produces stability in a different fashion than did the matrix of traditions and taboos which had ordered tribal society. Like taboos, laws contain negative injunctions that prohibit and threaten punishment for certain actions. However, that is the limit of their abilities. Forces that promote positive activity, such as the traditions that impelled the Blackfoot aspirant to care for widows and orphans, are absent. The negative cast to laws expresses the realization that the unfettered urge to dominate is suicidal. But laws cannot incorporate the encouragement to mutual aid that shaped tribal life. Such positive forces must arise from a different quarter.

In their village, the peasants experienced equality and connection with their fellows. They evolved customs that prompted mutual aid in any area that did not generate conflict with the aristocracy. The nobility, as they coalesced into a class, also developed traditions of cooperation to promote the well-being of their group. And the freemen in the towns generated their traditions of mutual aid in their communal organizations such as the guilds. These customs and traditions within the three separate groups generated some of the missing positive actions. But these three groups were inherently hostile to each other. What benefitted one could disadvantage another. There was no overarching loyalty which would be able to generate positive acts in any space that was shared by all three. A selfless action where the benefits were not restricted to their particular group, would inevitably benefit "others." Common spaces came to be devalued, treated as resources to be plundered lest any benefits accrue to those in another group. This division of the loyalties within society crippled it by making many essential positive actions impossible.

This is to be expected. We became Cooperators when we formed coalitions within which we could be safe. Care for this group is the basis of the positive traditions. However, these coalitions always drew a firm line between those inside the group and those outside. As we strengthened fellow feeling within, we accentuated hostility towards those without. When the tribe split into elites and peasants, and then when we added a third group of freemen who saw themselves as distinct from the other two, hostility amongst them was the complement to the growing mutual aid within. Competition for resources was a constant goad to this simmering conflict. The inevitable result was a total lack of traditions to spur individuals to nurture the community as a whole.

There have, however, been a few instances when a society was able to emerge from feudalism with a large degree of equality instead of experiencing the separation into warring classes. When everyone feels themselves to be an integral part of the same world, our powers to cooperate are released from the straitjacket of distrust. In the two cases we will look at—the city states of ancient Greece and those of medieval Europe—we see a shared regard for the whole of society that effectively eliminated antisocial activity. Curiosity was unleashed as exploitation was curbed. Through these two examples, we can glimpse the possibilities inherent in a more equitable society.

In Greece, by the eighth century BCE, a feudal aristocracy had consolidated its control into large fiefdoms and the gap between the peasants and the aristocracy was large and growing. Things were unfolding according to the typical script of feudal development, with the only question seeming to be whether their society would descend all the way to slavery. Then events took an unexpected turn, as historian William McNeil described:

> An important change in military tactics checked the drift of Greek society toward the middle eastern pattern of polarization between a leisured aristocracy and an oppressed poverty stricken populace ... The hoplite—a heavily armed and armoured infantryman—replaced the horseman as the decisive element on the battlefield. When massed together into a phalanx, hoplites could prevail even against

> cavalry ... The phalanx comprised eight ranks, each so closely formed that every man's shield helped to cover the neighbour to his left in line... By the early sixth century, the charge of a well trained and well armoured phalanx, its thousands of men acting as one and keeping pace by means of the rhythmic shouts of the paean proved itself all but irresistible.[1]

Military dominance came to depend on the many acting in concert as opposed to resting on the few who were able to provide themselves with horses and armour. Defending the granary depended on fielding a large army of well-trained hoplites. The distribution of power in these societies suddenly and drastically shifted.

No one knows the details of the invention of the phalanx. It would not have arisen from aristocratic experimentation as their cavalry were already an effective force. Also, such an army that could reward massed peasantry would be discouraged, perceived as a threat. Probably, some peasants escaped into the mountains rather than accept the enslavement towards which they were being driven. Living as hunters, a group of them banded together to try to protect themselves against their former masters who were actively hunting them down. A few, with skills honed by hunting, and inspired to stand together by some experience where they had been forced to work together, formed a wall of spearmen to stand against a small group of horsemen. After an initial success, they would have created larger groups to fend off an increasingly angry enemy. Even without armour, the scheme proved effective. As word got out that freedom and safety could be found in the hills, more peasants would join them. More recruits produced a larger phalanx. Experimenting with the armour of their victims, they would soon fit themselves out more effectively. Eventually, this army became strong enough to be used for attack as well as for defence. The aristocracy found themselves embattled. Being unable or unwilling to co-opt such a large group, the nobility fought for supremacy and lost. They were forced to flee or accept new roles as more equal members of this new society.

Athens and Sparta reacted to these new military possibilities in very different ways. In Sparta, a victorious band of rebels simply displaced the nobility, continuing to dominate the peasantry and a mass of captured slaves. This introduced a much larger noble caste, each of which had to live more simply. The Spartan phalanx dedicated their lives and the resources of the city-state towards becoming the ultimate fighting machine. However, because similar military power could theoretically be attained by any group of peasants, this amazing army could never venture far afield without risking revolt at home. Continual vigilance and harsh rule was the result. Sparta contained two classes whose interests were totally at odds. The allegiance of every individual was to their class alone. No one identified with the society as a whole. The Greek wonders in art, literature and architecture did not issue from Sparta. The invention of the phalanx did not produce widespread equality—it just presented the opportunity for it to arise. In Sparta, that opportunity was not realized.

In Athens, however, the rebellion took a different course. As in Sparta, it relied on the phalanx for success. As in Sparta, the aristocracy was defeated. However, the successful Athenian rebels remained embedded in the peasantry instead of becoming a distinct cadre of soldiers who could replace the old nobility. This difference probably reflected the course of events. In Athens, the aristocracy was likely overthrown quickly with broad support and community aid for the rebels while in Sparta the overthrow probably relied on rebels who had to live a long time in the hills pursuing a protracted war. This extended time would allow them to feel separated from the peasantry as they forged a new warrior identity. The idea of the phalanx likely first arose in Sparta, refined over a generation during a long series of victories and losses. After the example of Spartan success, however, an effective phalanx could be generated more quickly in Athens. Also, the vision of Spartan success may have prompted the Athenian aristocracy to yield more quickly, bowing to the inevitable.

In Athens, much of the peasantry gained land and became farmers. Every male with sufficient wealth to supply himself with helmet, shield and spear was expected to join the phalanx and contribute to the defence of the city. This contribution to defence entitled him to citizenship. Most farmers could meet this standard. And since the restriction was one of means, not birth, any poor male could aspire to acquire the weaponry that would win him citizenship. When the army was not active, everyone went back to the farm, only getting out the shield and spear for practice sessions. As the city became wealthier, potters, merchants, vintners and boat builders could also afford to join the ranks of the phalanx. No separate class of soldiers existed. Athens was essentially defended by a militia. Most of the freemen in the polis were citizens, able to participate in decision making and governing as well as defence, members of a class with none above it. Slaves were numerous and often treated brutally, especially in the silver mines, but they were kept separate by force from the life of the city. Women were largely confined to the home and forced to experience the polis through their fathers or husbands. So, though equality was by no means complete, it had extended to a group numbering in the tens of thousands.

4. The Growth of the State

When the Persians advanced an army of conquest into mainland Greece, a strong navy became just as essential to the city's defence as a strong phalanx. Rowers became vital. The wealth restriction disappeared and citizenship spread further, soon encompassing most non-slave males in the polis. Equality, and the new political form of democracy which expressed it, was a reflection of their military configuration. Trust for those holding their shields in your line and for your phalanx which stood or fell as one was essential. There were not three classes with differing loyalties in Athens, or even two as had developed in Sparta. Athenian society was made up of only one dominant group, with all other people forced far from the public realm. All citizens identified as a member of the same group. There was no "other" in the city.

Each citizen of Athens, as well as being a worker and a soldier, was expected to take their turn filling the civic positions and participating in deliberations of the community forum. For everyone, being an Athenian defined his life. It provided identity, meaning and context for everything he did. This echoed the tribal past. As equality was valued, a simple style of dress became the norm. Conspicuous displays of wealth, emphasizing differences, were avoided. Wealth, when accumulated by individuals, was translated into esteem by enhancing the polis with public buildings, public performances or public art. Everyone shared in glory brought to the city. Everyone shared in any shame. This identity of the individual with the polis restricted actions which would be harmful to the community.

Athens was the richest and most successful of the many city-states which arose at this time. Since the wealth from its slave-operated silver mines allowed it to build the strongest navy, it assumed the lead in the coalition that formed to repel the Persian invasion. At the successful conclusion of this campaign, she found herself dominant within this group of cities. The strong sense of equality which existed within her borders did not extend to these others. They were not people with whom the Athenians wanted to share. The urge to dominate these people was not prohibited by custom or taboo. Largely because it helped Athenian merchants exploit weaker cities, the league and the navy were kept together after the war as the Athenian empire. Athenian citizens became privileged in the Aegean. As trade flourished, more workers were needed in Athens. The city grew by acquiring slaves and permitting non-citizens to live within her borders. Actions that enhanced the polis now benefitted many deemed unworthy. The identification of citizens with their polis ebbed, replaced once again by a narrower allegiance to a specific class.

Athenian farmers, assured of a supply of grain from elsewhere in the empire, turned to the production of figs, olives and grapes, crops more suited to the rocky slopes of Greece and more valuable as commodities for export. Trade in these items brought ever more wealth to the city and spurred the development of other industries such as shipbuilding and the production of amphora. Many Athenian farmers became so wealthy that they hired mercenaries to fulfil their obligations to the phalanx, adding the professional soldier to the mix in the city. Soon citizens were just a privileged minority amongst a mass of non-citizens, slaves and mercenaries. Generosity towards the community ceased. To enrich oneself at the expense of the polis was no longer unthinkable. Taboos had to be reimposed to protect the common spaces, people and resources.

But for the few generations that people fully identified with the polis, they produced a culture that was amazing. To be seen as the author of an antisocial act was the ultimate shame. People lived their lives according to common ideals, a common vision and trust in each other. Each person was his own master; he alone could decide when extra hours of work were worth more than time spent elsewhere. Freedom, leisure, extra resources and the ready esteem of the group unleashed the Originator. Many turned to science. Socrates, Plato and Aristotle turned their curiosity towards the meaning of life. Aristophanes, Sophocles and Euripides created drama. Herodotus and Thucydides invented the writing of history. And thousands of others, some remembered, most forgotten, turned their creative eyes on amphora decoration, architecture, building, sculpture, painting, oratory, writing, acting, boat design and every other aspect of their lives. The works that emanated from those few thousand people during that hundred year period astound the world today.

If the Peloponnesian War had not led to the downfall of the democratic polis in Athens, some other event would have accomplished the task because democracy was seeping away as equality disappeared. As H. D. F. Kitto, the English historian, compared the citizens of the time of Demosthenes and the Peloponnesian war with those of the age of Pericles and the Persian war about a hundred years earlier:

> The contrast between the age of Demosthenes and that of Pericles is startling; to Periclean Athens, the idea of employing mercenaries would have seemed the denial of the polis—as indeed it was... What we meet in the fourth century is a permanent change in the temper of the people; it is the emergence of a different attitude to life. In the fourth century, there is more individualism. We can see it wherever we look—in art, in philosophy, in life ... sculpture ... begins to portray men, not Man. ... The ordinary

citizen is more interested in his private affairs than in the polis ... Demosthenes compares unfavourably the splendid houses built by the wealthy of his own time with the simple ones with which the rich of the previous century had been content. ... The leading figures in the assembly are no longer the responsible officers of state too. Still less are the responsible officers of state also commanders in the field.[2]

The meaning of citizenship was transformed by the opportunity to dominate. Power corrupts.

The Greek city-states were incapable of cooperating with each other. It seems that the stronger the internal sense of pride, the stronger the corresponding distaste for the "other." Philip of Macedon, an aristocratic ruler who based his own army on a phalanx of mercenaries and conquered peoples, managed to unite the Greeks by defeating them one by one. His son, Alexander the Great, led these united Greek armies against Persia and conquered the Middle East, Egypt and well into India before his early death. Soon after this, the Romans created another military machine, refining the phalanx still further until their army became an unstoppable force. A small number of aristocratic Roman citizens directing a large number of legionnaires from the peasant class enforced Roman hegemony over a massive territory. They crushed all attempts at local independence within most of Europe, North Africa and the Middle East for seven hundred years. By the time the Roman Empire collapsed, advances in saddles and armour had swung the advantage in arms from massed infantry back towards horsed knights. The successor to collapsing Roman power was once again a world where horsed knights dominated peasant villages to rule a feudal estate. This patchwork of fiefdoms embarked on the inevitable and violent process of consolidation toward statehood. By 1000 CE, Europe was ruled by a group of warlords groping towards becoming an aristocratic class.

Towns were on the rise. These were walled for protection, housed industries which provided employment, and offered freedom to peasants fleeing the estates. As they grew, they forged connections with each other for both protection and trade. The most successful of these were the Lombard League in Northern Italy, and the Hanseatic League along the northern littoral of Europe.

The wealthier merchants made a grab for freedom. They armed their workers and claimed independence. The lesser nobility, sensing the winds of change, attempted to align themselves with the mob to wrest a stronger position for themselves within the aristocratic hierarchy, but their efforts were generally too little, too late. In most cases, the merchants and the guilds had become too powerful to control. In the northern Italian cities, popular uprisings first installed aristocratic committees which were soon replaced with more broadly-based governments. The city-republics had appeared.

These cities grew rapidly once they escaped the restrictions of feudal control. Peasants streamed in from the countryside. Trading expeditions were soon venturing around the Mediterranean and beyond, principally from the cities of Genoa and Venice. Cloth manufacturing became a major industry in Florence. The metalworkers in Milan became numerous, renowned and profitable. Some of the families leading these industries—the Medici, the Borgias and the Sforzas, for example—became very wealthy. A similar growth in trade fuelled an increase in wealth in the Hanseatic cities.

Townspeople created the guild structure to replace the benefits they lost when they abandoned the estate.

The feudal world guaranteed them food and protection in return for their subservience and labour. The guilds filled this gap with militias and courts for protection, aid in times of sickness and troubles, guildhalls for conviviality, and history, traditions and a pantheon of heroes to show themselves who they were. When revolution arrived, the militias were already cohesive units with a ready source of organized muscle. The townsmen knew what they were protecting. They went to war shoulder to shoulder with their fellows, highly motivated to retain their freedoms. The mob already had an effective scaffold of organization.

Each guild member was fighting for full citizenship, and he expected it to be his after victory. However, the richer merchants, once accepted into the ruling circle as representatives of this power, were able to advance their position at the expense of general enfranchisement. Effective control in most towns was exercised by a group that only numbered in the hundreds. The economic might of these merchants coupled with the military power of the lesser nobility provided stability.

The ruling groups, though more limited than what the workers may have wanted, had been greatly expanded. Positions in the structure were shared by up to five hundred regular holders of seats in the various councils. And significant power did still reside in the guilds, resting on their being the source of the militias and a respected place of connection, trust and mutual aid for each worker. Power no longer rested exclusively at the top—it had spread quite far throughout the guilds and the councils of state. Each organization had a large degree of autonomy, with negotiations being necessary wherever their interests overlapped.

During the period of struggle, people believed the city belonged to everyone and not just to its rulers. This reversal of the feudal mindset informed the worldview of many members of the councils as much as it did the workers. Many of these rich merchants had been workers themselves not so long before. Though there was distance between classes, it was not nearly as great as the gulf that had existed between peasants and aristocracy. Also, for the workers, the removal of feudal restrictions fanned the flames of personal pride, guild pride, civic pride and identity with the city. A Florentine or a Venetian, from the richest to the poorest, was as much defined by the attributes, fame and opportunities of his city as was the Athenian by his polis or the Blackfoot by his tribe. For some time after the revolutionary period, even as class distinctions were beginning to grow more rigid again, this identification of the individual with his city persisted, both for freemen and councilmen. However, the actual disparity of power provided a constant opportunity for the more powerful to tighten their grip on power. The political system tended to drift towards despotism or oligarchy.

But for a short period, everyone identifying with their city meant that much of the wealth produced went to enhance the community. Bringing glory to the city gained the ultimate esteem for the individual. Patronage of the arts and public architecture was expected. The Originator once again turned his curiosity onto history, science, government, economics and philosophy. As writer and social critic Paul Goodman summed up:

> In fact, these towns, associations of associations, some of which were hierarchical, some democratic, invented modern technology, founded the modern university, hit on modern corporate law, and, by and large, created modern civilization.[3]

Just as had been shown to be the case in Athens sixteen hundred years before, when everyone identified strongly with the whole of society, excesses are reined in and energy is channelled into positive, creative pursuits.

However, some merchants were able to use their wealth to continue to enhance their political power. In a few generations, an expanding gulf appeared between elites and workers. Allegiance to a class replaced the allegiance to the whole of society. The wealthy changed their goals from public glory towards an appetite for luxury. Wealth again allowed the hiring of mercenaries who would answer to the leadership more reliably than did guild militias. It was the old story. The cities started to fight amongst themselves for control of the food-producing countryside and the trade routes. These wars drove the citizenry deeper into poverty. In his book on the Renaissance Italian city-states, Professor Lauro Martines echoed the comments of H. D. F. Kitto:

> The Quattrocento building and rebuilding craze was the very process of elite consciousness redefining itself in terms of new needs, new pursuits, new satisfactions ... They sought to affirm themselves by means of more imposing palazzi, more organized and more splendid facades, wider and higher internal spaces, a display of finer manual work of all sorts, a higher finish to things, more rounded edges and polished surfaces and larger accumulation of objects ... They accorded perfectly with the sharp rise of fancier and fussier styles of dress which required finer handiwork ... And the changing social identity profiled above was one ever more intent on its authority and forms of display; it was turning more worldly, more refined, more ostentatious, more boastful.[4]

The identity that common workers felt with their cities disappeared. A despot was a despot and they ceased to care much who their overseer was. The elites owed their positions to judicious hiring of mercenaries and this entailed obvious risks when there were other bidders for those services. The city-states were attractive to attackers because they were rich and could produce wealth. They fell, one by one, and were reabsorbed into empire. The workers were indifferent—they just had to keep their heads down until the change in masters was complete.

Equitable societies induce individuals to identity with their society. This unleashes civic activity that cannot appear in a society of separate classes. If everyone in your life shares common hopes, fears, pride, shame, responsibilities, problems and solutions, then destructive activities disappear while constructive pursuits become reflexive. The Greek city-states let off an intensity of genius that outshone Alexander and Rome. The European city-states produced the intellectual capital that animated the Renaissance and subsequent western development. These two glorious episodes are beacons in the sorry chronicle of feudal history.

These two brief episodes, however, were notable for being exceptions to the rule. Elites generally maintained control as their kingdoms grew. They built a mythology and a working set of laws, and they gradually coalesced into an aristocratic class. Those first protectors of the granary widened their reach until they sat astride a state. As the world filled up with states, they started to interact in the thousand places where their interests touched. It was time for the elephants to learn to dance.

CHAPTER FIVE

The Inevitability of Empire

∿➤

No matter how inspired we may be by classical Athens and the medieval European city-states, these were exceptions to the inexorable historical process. The human drive to dominate impelled soldiers to create kingdoms ruled by an hereditary elite. They firmed up their borders with treaties, established laws and created a story to explain their position. The result was a state. But laws and treaties did not alter the urge to dominate that drove these men. They had the wealth to arm themselves. If they could find victims to exploit, it was sure to happen.

There were places in the world with whom they shared no borders and had signed no treaties. States did not launch imperial forays in order to create secure borders or to find a needed food supply. These needs had been largely met before they generated a state. The reason for all imperial adventuring was a search by the Dominator for plunder and victims. People being what they are, states are by nature imperialistic.

Throughout recorded history, states have attempted to project power beyond their borders. This always involves risk. Support for an imperial army leaches wealth from the state inviting either conquest by neighbours or rebellion from within. Failure to win sufficient booty in an imperial war creates a net loss which raises these risks considerably. But success also leads to problems. A rich prize tempts challengers. The more distant and wealthier the prize, the greater the effort and cost required to defend it. Much of the booty, so glittering on seizure, can easily be swallowed up maintaining control over a colony. Capturing an empire can be the easy part. Holding it together demands continuous war and the cost of this is never clear in advance. Conquests can easily turn into millstones. And should the well-being of the imperial state come to depend on the spoils of empire, then every challenge becomes an existential threat. History records a succession of empires that rise, expand through brilliant military campaigns, loot resources mercilessly to create opulence at home mirrored by poverty abroad before imploding into chaos in spectacular fashion.

But between those end points of conquest and collapse, an influx of imperial wealth can help an aristocratic class to define itself. These stolen resources can fund monumental homes, lavish fashion, extravagant living, and scientific, literary, artistic, sporting and exploratory pursuits designed to win esteem from other members of their class. These accomplishments and displays bolster class pride and justify their view of themselves as superior in the world. In Athens, during the few centuries after the classical age had turned into empire, while mercenaries were policing the seas and slaves were working the orchards and mines, much of the wealth of empire was still funneled into such pursuits.

The only certainty following the growth of an empire is its collapse. It is the tragedy of Athens, the ruins that now define Rome, the appalling environmental destruction of Mesopotamia. Chinese history is a saga of kingdoms consolidating into imperial dynasties interspersed with

5. The Inevitability of Empire

violent fragmentation into competing warlords before a new pretender rises from the rubble. India shows a similar picture of local rajas consolidating into warring kingdoms, only to find themselves subsumed one by one into a foreign empire—first the Mughals and then the British—who took advantage of the animosities between them to gain control. The Inca swept through the Andes subjugating all other tribes only to have the Spanish arrive and take advantage of the disorder of a moment of dynastic competition and civil war. Struggles to establish and hold empires are the chapter headings in the story of feudalism.

When soldiers looked outward for victims that they could subsume into their empire, they saw a variety of peoples ranging from nomadic tribes to isolated farming villages to militarized states similar to themselves. It was obvious that states similar to themselves would be difficult to conquer. They could be raided for plunder and conversely, they were lurking threats requiring armed readiness. But they could only be challenged by engaging in a huge war which could very well end in defeat.

Collections of small farming demesnes that had not yet consolidated into a militarized state were helpless in the face of an imperial army. They could be picked off one by one and the peasants either put to work for the conquerors or sold as slaves. Local leaders could be killed or co-opted. This is the Incan empire. Farmers could continue their lives only when they were useful as workers.

A nomadic tribe was never so lucky. They were not farmers, so could not work land for a new master. They did not have the accumulated wealth necessary to support a standing army to oppose the invader, no matter how clearly they might discern the threat. Their military traditions were inappropriate, their weapons were technologically inferior, and the numbers that hunting can support always left them outnumbered. In a battle between a tribal culture and an imperial state, the tribes could only be victims.

The prize was the land. A territory can support game for hunters or it can produce crops for farmers. It cannot do both. After the battles were over and tribes were reduced by war deaths, starvation, massacres, enslavement and disease, the land could be transformed into fields for the empire. Defeated tribesmen had no place in such an agricultural order. The cultural gulf is too wide for nomadic hunters to transform themselves into estate labourers. Instead, victors would import colonists with the appropriate expectations and skills to work the farms. If hunting tribes were able to migrate onto land that was ill suited for agriculture, then they could continue to survive. But without access to such a retreat, they would be exterminated and separated from their land. The North American tribal wars where small groups of warriors armed with bows, knives and spears faced detachments of professional soldiers wielding repeating rifles and cannon is an example of such a war. The conquerors saw tribespeople as so "other" that they felt little compunction in genocide. War between a state and a tribal people was waged for land that could be settled by the priests, merchants and farmers trailing in the wake of the army.

Occasionally, a reprieve could occur if hunting could produce more value than the land itself. For example, in Canada, the Plains people survived for a few generations as they delivered valuable furs from land that would be difficult to farm. This situation was only temporary but a few generations of contact and trade allowed some degree of adaptation to the new cultural norms. An assimilated place in the new order could possibly then be found, people surviving even though still being obviously the "other." As more and more land fell under the plough, tribes were confined, assimilated, killed or forced to move on. A nomadic hunter has no place in an agricultural empire. He is only an echo of the distant past.

In the modern age of exploration, the spoils of empire were foodstuffs, spices, slaves, furs and precious metals which, when shipped to the heartland of the empire, became luxuries for the elites. But there were times when such plunder had no immediate use. More slaves are not useful if there is not more menial work to be done. More metals are not useful if you already have as much as your industries can use. Food is useful when captured but may spoil before reaching home. The push for empire continued unabated, fuelled as it was by the urge to dominate but, in many cases, the conquerors found that they were better enriched by finding someone with whom they could trade immediate plunder for durable wealth that could be accumulated. The travelling merchants that had visited their castles for hundreds of years had demonstrated the concept. Indeed, there were merchants eagerly following the conquest looking for just such opportunities. An empire with far flung outposts, each with their own particular products and needs provided ample opportunities for trade. Booty could be diverted for quick advantage instead of sending it all the way home.

As had happened with trade in the past, promissory notes and scrip of various sorts greased the wheels of international trade. Banks developed new services and a wider footprint to make this trading safe and profitable. Both the military and the financial adventurers hoped to retire rich, using this wealth to push their way into the ranks of the elite.

Trade stimulated manufacturing at the heart of the empire by providing new markets and new sources of raw materials. This was vital to the wealth of both Athens and pre-industrial England. Population in the motherland could be encouraged to grow without limit as excess people became colonists: both the younger sons of aristocratic families and unemployed workmen sailed off to populate newly-conquered areas.

Soldiers pacified new territories. Merchants supervised the removal of assets and fed this plunder into a world of international trade. Indentured colonists arrived to exploit local resources. Profits flowed into the bank accounts of the aristocrats and, increasingly as time went on, into those of the merchants.

In the medieval European cities, merchants had accumulated enough wealth to finance trading expeditions without the aid of the elites. Groups of Genoese and Venetian merchants, roaming the Mediterranean to buy and sell in far flung colonies, distant empires and other city-states, became rich and powerful. Flemish merchants did the same as they turned the ships that had successfully carried goods between the Hanseatic cities towards the Volga, the Atlantic and the North Sea. Wherever strong royal prerogative had prevented merchants from taking root, such as in Spain, the nobility planned, built and controlled ships and exploratory missions as national imperial adventures. As the modern world began to take form in the sixteenth and seventeenth centuries, reliable profits were available for merchants in Caribbean sugar, Canadian furs, African slaves, American cotton, and spices, cloth and teas from the Far East. National naval protection allowed this trade to increase rapidly. This wealth was added to the profits these merchants already commanded from fabrics, metallurgy, distilling, brewing, boatbuilding and other industries which were thriving under the twin stimuli of colonial demand and aristocratic consumption.

The aristocracy were a closed class which shunned these upstarts. The merchants, with their expanding wealth, had risen far above the workmen and artisans so that group too was foreign to them. As competitors for investment opportunities, they fought with and distrusted each other, making the formation of a cohesive merchant class initially impossible. Individually, they had great power and great wealth but they were as isolated as were the first soldiers that had stood above their tribe. Denied the context of an encompassing group, they doubled down on dominance as their route to meaning and satisfaction. The wielding of wealth could destroy a competitor as surely as if you stormed his castle and stole his grain. Economic battles also produced booty and bootlickers. Wealth is power. These merchants competed fiercely for the best opportunities to reinvest their accumulating wealth. Driven solely by the urge to dominate and to hoard, they each strove to get richer in whatever way they could. This was how they expressed who they were. Their satisfaction arose from the winning of money and power, from expressing domination in the economic arena, not from the actual wealth itself. No amount of money was ever enough.

These massive flows of money into the European cities created the Industrial Revolution. Rich investors needed more goods to trade than piecework was providing. This led first to the harnessing of falling water to drive industrial looms. With the quick development of good river sites, other investors turned their attentions to other ways to power such mills. The basic properties of steam were already known and Thomas Newcomen, James Watt and a host of other tinkerers were rewarded by investors as they solved the various problems that impeded the construction of an effective steam engine. This permitted mills to be located wherever there was a source of coal and access to ports. Money from other investors financed canals which allowed mills to be built increasingly distant from both coal and ports. Within a few decades, locomotives made most of these canals obsolete. The accumulating capital from wildly successful international plunder, all of which needed profitable investment, was the motive force which called the Industrial Revolution into existence. These mills, in turn, required a large work force within walking distance of their doors, and this unexpectedly redefined the political structure of the world as it raised the spectre of the mob in entirely new ways.

All empires in history have followed a similar path from tribe to village to kingdom to empire. Merchants were present in every empire, following in the wake of the soldiers, but, as with the Spanish example, control had always remained firmly in the hands of the nobility. The plundered wealth had exclusively been directed to finance the activities of the aristocracy. In preindustrial Europe, however, free merchants were able to claim much of the profits and they were mainly interested in directing their accumulating wealth towards reinvestment instead of display. Impetus for change began to come from a non-aristocratic class. As riches begat riches, they became ever more wealthy and influential. Merchants were present all over western Europe, but the process went furthest and fastest in the British Empire. To understand how industrialization shaped the modern world, we will now take a short step back in time and look briefly at the story of the British Empire.

When the occupying Roman armies were recalled to help defend Rome in the second century CE, the England they left behind lapsed into a collection of feudal farming areas, each controlled by a local strong man, and the familiar script of consolidation began to unfold. Seven hundred years of battles coupled with steady Norse colonization off the North Sea had, by the turn of the millennium, produced three large kingdoms, each of which had made alliances when necessary with kings in Normandy and Denmark to support their claim to rule the whole island. This muddle produced a variety of claimants to the throne, one of which was William of Normandy who advanced his claim by bringing over an army and defeating a coalition of Anglo-Saxon nobles who had just been weakened by repulsing another invasion force of Danes in the north. The nobles were replaced in their castles by Norman retainers loyal to the new king. The peasants kept their heads down and continued to till the fields. England was unified.

William had seen how the king of France had been continuously thwarted by the strong regional power bases of his retainers. Though they swore fealty to the king, they were such powerful rulers over subsidiary territories, of which Normandy had been one, that they could choose to obey or not. To avoid this situation, William distributed lands such that every noble got his wealth from land dispersed through the kingdom rather than from one contiguous demesne. Insubordination became more difficult. England was functionally unified a full 500 years before France and a good 800 years before Germany.

Because of the extensive coastline, the British had always ventured to sea, developing skillful sailors and sturdy vessels that could survive the open ocean. The Norse technology of oared ships had added to this base of knowledge. When the newly consolidated Norman aristocracy looked outward for victims, the means to reach them were at hand tied up in their ports. The Crusades provided an early outlet for restive soldiers. Recapturing the resting place of Jesus was the avowed goal but establishing kingdoms in the Levant was always the real prize. This futile endeavour was followed by the Hundred Years' War, a century of sporadic slaughter and pillage as English knights tried to establish and hold kingdoms on the continent in the face of equally powerful adversaries. The cost of this folly virtually bankrupted England. These adventures made it clear that, if possible, the search for victims should be directed at much weaker opponents. So they joined the competition between the Spanish and Portuguese to sail farther afield and find victims along the African coastline, in the Caribbean Islands, the Americas and the islands and coastlines of the Far East. They were looking for opponents that were weaker and more primitive in weaponry. They found such peoples, conquered them, founded colonies and opened these to trade by their merchants. This trade grew steadily. The merchant, banking, administrative and entrepreneurial class who controlled it expanded in numbers and wealth.

By the seventeenth century, the merchant class was strong enough to align themselves with the minor nobility and challenge the king and the elites for power. The few generations which saw the English revolution, the Cromwell Protectorate, the restoration, and the subsequent "glorious revolution" of William and Mary, reoriented the government to one which was more sympathetic to the needs of the merchant class. The old aristocracy secured their continued existence by agreeing to share some of their power with these upstarts. This new government had very different priorities than the old. As economist Harry Magdoff wrote:

> The English revolution of the seventeenth century ... put its stamp on the whole era conditioning the rise of Britain's leading role in empire, finance, and trade ... Under Cromwell, Britain set out to build for the first time a national and professional navy ... The Navigation Acts of 1650-51 ... not only created a

monopoly for Britain's ships in its trade with Asia, Africa and America, but also created the basis for a whole set of restrictions on its colonies which gave an important boost to the demand for British manufactures.[1]

There were financial gains in this new world for the landed aristocracy because many of the industries were closely tied to the land. Woolen mills required land to rear the sheep. Coal and other minerals were mined on someone's estates. Shipbuilding required stands of timber for which someone had to be paid. The growth of towns provided rents and real estate windfalls to the nobles on whose territory the town stood. Breweries, milling and food production all required access to land, producing payments without the aristocrats having to actually engage in business themselves.

The aristocracy had never been a completely closed caste. Ennoblement had always been dangled as a possibility. Merchants who made fortunes could aspire to eventually be admitted into this upper class. By educating sons in aristocratic schools where they could assimilate the ethos, the accent and the manners of the ruling class, by currying favour, by buying estates to look like nobility, by rescuing failing aristocratic families with an ennobling marriage, it was possible to go from rough merchant to aristocrat in a generation or two. The pageantry, the colourful trappings, the self-centredness and self-confidence of the elite constantly reinforced the idea that this was the only worthy goal. Since some merchants accepted this goal, and since the strongest among them periodically abandoned their ranks to become part of the elite, the merchant class did not challenge for complete control. Also, occasional infusions of new blood and new money helped to keep the nobility connected to the changing times. Assimilation helped the aristocracy maintain some measure of control for centuries after defeat in the revolution.

Manufacturing grew, trade multiplied and businesses boomed as the empire expanded. As merchants became richer they needed ever larger investment opportunities. When the steam engine appeared, constructing factories was seized on as a way to multiply their capital. The stumbling block was that large factories required plentiful labour as well as abundant capital. They needed people who would work cheap and endure terrible conditions. If people were given a choice between life in a farming village or existence in a mill town, the factories would remain unmanned. One solution was the Enclosure Acts which made traditional life in the country no longer tenable for many. As journalist Felix Greene observed:

> ... the landowners found raising sheep more profitable than renting land to tenants. Thousands of peasant farmers were evicted from their cottages, uprooted ... from the land that they and their fathers had used from time immemorial. What caused even more widespread suffering were the Acts under which public or 'common land' were enclosed. In accordance with age old tradition, all men were free to use these common lands for the grazing of sheep and goats; in the economy of the peasant farmers, access to this land was an essential element without which they could not survive. Between 1760 and 1810, no fewer than 2765 Enclosure Acts were passed. Thus it happened that, when the new factories that were springing up required labour, tens of thousands of homeless and hungry agricultural workers with their wives and children were forced into the cities

in search of work, any work, under any conditions, that would keep them alive.[2]

With acts of parliament, genteel acts of speech and pen masking the inherent violence, fleece was made available while desperate workers were forced into the slums of the industrial cities. The industrial working class reluctantly appeared in the pages of history.

These workers were paid wages that were barely sufficient to keep them alive because they had no other option but starvation. Children also worked and were paid even less so long as a family's total wage packet would keep them alive. Removed from the land, the peasants lost the ability to augment their wage with personal growing and gathering. Where they may have been able to harvest materials to improve their home, this was no longer possible. Where they may have been able to garden, graze animals, hunt and fish, these options also disappeared. The completeness with which a ruling class controls the food supply always determines how harsh the rule becomes. Control was complete. The rule was ruthless. The workers lives were insecure and brutish, barely above slavery.

The extreme power imbalance meant that wages were kept low, and this led to huge profits. Investments in these new mills were vastly and quickly repaid. At the same time, capital invested in overseas trade also returned large profits. For example, the East India Company paid a regular dividend of 10 percent to well-connected investors. The Hudson's Bay Company generally returned between 10 and 25 percent throughout the mid-eighteen hundreds. Fortunes did not just grow, they multiplied. In this short period of explosive growth, you would have to be an idiot not to make a fortune if you had capital to invest. These gains also needed to be invested, so every success pushed the whole system to grow. More mills were built. More colonies were founded to provide resources and markets. These merchants were driven to invest—to increase their holdings and their income. There was no other source of satisfaction available to them. Their goal was the power and eminence that wealth conferred, not the money itself. They were propelled along this path of furious expansion by the unleashed Dominator in every merchant breast.

Unlike traditional imperial growth which is nominally under the direction of a king, there was no overall planner in a world created by investment. The basic rules of supply and demand shaped the competition and they had all accepted a few legal limitations, but there was no blueprint directing their activity towards some vision of the future. The impetus to grow and the direction that this growth assumed issued from myriad independent decisions, all shaped by a shared urge to dominate.

When the capital available is less than the investment opportunities, as was the case at the start of the Industrial Revolution, then opportunities are plentiful and profits are large. Everyone can win. Such a burst of growth, persisting for less than half a century, convinced the merchants that their success was due to their genius and daring and that they would continue to win no matter what they chose to do, no matter how underlying conditions might change. This arrogance caused them to discount risk, a stance which compounded their gains when times were good but left them open to disaster when growth stalled.

The hazard in such unrelenting growth is that the world must eventually become saturated with your products. At that point, further investment is obviously unwise. However, profits from past investments continued to accumulate. Investors, locked in their competitive struggle, continued to invest. Competition intensified for even the least-promising opportunities. Merchants built new mills in answer to accumulating capital, arrogance and self-deluding hope—hope that the glut is somehow temporary, hope that their product will fare differently from everyone else's, hope that they are so brilliant that the laws of economics do not apply to them. The urge to dominate overruled logic, just as it had with the early soldiers.

Their actions increased the glut of goods and drove prices lower. This guaranteed that new investments would fail, while making all mills less profitable. Marginal mills—those less well organized, those whose coal must be brought from farther afield, those paying too much interest on borrowed money, those with any weakness at all—failed, ruining their owners. But these shuttered factories were not dismantled. They were scooped up at bargain prices by vultures with deep pockets who were prepared to wait. When enough plants were idled that prices started to rise, the closed mills restarted, driving everyone back into unprofitability. Getting beyond this glut could only be done by permanently expanding the market. Since the producers would not stop growing, more buyers had to be found. British history since the start of the Industrial Revolution has been a search for markets to match the exploding capacity.

Domestic markets were limited. Workers could scarcely keep themselves alive let alone purchase consumer goods. The aristocracy and the small but growing middle class of artisans, shopkeepers, bankers, lawyers and administrators did have a limited need for more blankets, silverware, whiskey and bullets, but this market was not growing nearly as quickly as the factory system itself. Hand production had been adequate to meet the needs of this group before the mills multiplied capacity. There had been no shortages. The Industrial Revolution was a response to capital needing investment, not to shortages of goods or unfilled needs. External markets became essential. It is painfully ironic that the Industrial Revolution created a glut of goods at the same time as it caused exceptional hardship in the lives of the workers.

Sales agents were dispatched around the world to sell these goods. The British navy had become dominant on the oceans and would open and police any markets that her traders demanded. This was crucial to the continuance of industrialization in England. Other European countries had capital, coal, access to steam-driven machinery and peasants to exploit, but they lacked England's ability to subdue and control overseas markets. As in England, their manufacturers started down this road of expansion; as in England, markets became glutted, profitability shrank, recessions emerged, and bankruptcies followed. The bargain price assets, however, could not be rescued by expanding into new markets around the world. Growth being impossible, the industrialists would continue to bleed financially as their mills fell into ruin, weakening the merchant group vis-à-vis the aristocracy. Recessions just left the European aristocracy more dominant. But for English merchants, trading in the wake of the navy could keep the party going. The fortunes and the power of the English merchant class steadily increased for the better part of a hundred years.

A navy that could not be opposed and an army that was prepared to go anywhere in the world to advance the interests of English merchants permitted them to write their own terms. Between 1815 and 1914, there were only fifteen years when the British army was not engaged in colonial wars to protect and advance British interests. Forcing harsh terms on the workers at home, and trading abroad under similarly enforced profitability, the English merchants' fortunes grew impressively. This growth, though described as the result of free trade, rested solidly on the willingness of the British military to do the bidding of her traders. Imperialism had become a little more subtle than massacre and pillage but it was still driven by the need to dominate and it still rested on the use of force.

This unchecked expansion of fortunes, factories and foreign trade for more than a century transformed British society. The millions of workers that had been crowded into the cities became a new and potentially explosive force. Because of their numbers and their desperation, they were a credible threat to both the aristocracy and the merchants, as the example of the French revolution had made abundantly clear. Far from the dominating presence of the lord on the estate, these people were able to question their situation. The exhilaration of gathering in large crowds gave them a hint of their power. The merchants' urge to invest and the army's success in making markets expand had produced the industrial working class.

The middle class, neither part of the working class nor accepted as equals by either the aristocrats or the industrial magnates, made up of artisans, doctors, teachers, clerics, lawyers, shopkeepers, and managers in the factories, trading houses and banks, had also grown significantly as the industrial world created ever more opportunities for them. These people saw their well-being tied to the success of the industrial system so they were not perceived as a threat by the ruling elite. However, as they became more numerous, they too were able to press demands for more privileges. Since their work was essential to the profits, these demands had to be taken seriously.

And then there were the merchants, a small group with power disproportionate to their numbers. They were no longer controlled, as were their forebears, by hopes of joining the aristocracy. Some were unacceptable because of their origins, their attitudes, their religion or the way in which they had made their money. Most, however, had just come to realize that, as their numbers grew, ennoblement became less of a possibility. The aristocracy could absorb only a few new faces in any generation without diluting their own identity. The group of magnates was now so large that this hope was not realistic. These men were wealthy, brash, confident, independent and driven to dominate. They chafed at traditional restrictions which asked them to defer to the aristocracy. Through their wealth, they controlled many key aspects of society.

So there were now four powerful elements in society—the industrial working class, the educated middle class, the industrial magnates and the aristocracy. The power of the first three groups was increasing year by year, while that of the aristocracy was in stasis if not in actual decline. The rural peasantry, though still numerous, was so dispersed as to be powerless in this new world. The aristocracy controlled the land, the parliament, the army and much of the legal process. They captured some of the wealth of industrial success through rents, but their ability to compel either the working class or the magnates was no longer a given. Traditional cultural norms, which had evolved to define a relationship between an aristocracy and a dispersed peasantry, successful in ordering society for centuries, no longer guaranteed stability.

From the late 1700s until about 1870, imperial expansion powered profitable growth which created opportunities for the middle class and wealth for the industrial magnates. Both groups grew in numbers and in power. The industrial worker, though still living in poverty, increased their numbers dramatically. As they coalesced into a cohesive class, the inherent threat of mob violence they presented forced their demands to be taken seriously. Concessions were reluctantly ceded to them by employers and through the legal structure, on wages, safety and hours of work. Their lives slowly improved. As long as the system was expanding, these benefits did not have to come at the expense of any other class.

Around 1870, however, expansion hit a wall, not just in England but over most of the industrialized world. It had become impossible to find new markets in the volume that were required. New and growing competitors, such as the United States and Germany, were also seriously industrializing. Efforts to protect markets against such strong competitors could lead to war on a scale that Britain could not undertake lightly: these were states with technologically advanced weaponry. The underlying problem was that the easy markets in the world had been found, colonized, opened to trade and exploited. But capital continued to accumulate. The great recession that defined the late nineteenth century settled over the developed world. Investors continued to invest as they were driven to do by their arrogance and their irrational hopes. Gluts piled up. Prices collapsed. Businesses failed. Unemployment soared. People starved. The good times that had rested on expansion came to a crashing halt.

A large pool of capital desperately seeking investment opportunities is a powerful force. Like water building up behind a dam, it will seek any outlet to relieve the growing pressure. This pool of capital powered the second industrial revolution by providing large inducements to innovators who could bridge the gap between academic science and new industries. Scientific knowledge had been growing in the wealthy societies but that knowledge had remained largely theoretical as long as investors could attain safe profits with investments in industries with which they were comfortable. However, when desperate investors offered handsome rewards, scientists and engineers changed their focus from gaining understanding to crafting practical applications. This resulted in a range of new products and services that were based on steel, oil, electricity and industrial chemicals. By the 1890s, these new industries were taking off, providing uses for capital on a large scale. Whole new economic sectors appeared, literally overnight.

This transformation accelerated changes in the power relationships among the four classes in society. These new industries brought new people and new profits to the magnates. The numbers and the power of this group increased dramatically. It also led to explosive growth in the educated middle class because these new industries were more complex, more based on knowledge and organization than the old. The working class was essential to man these new factories, so their numbers also expanded. And because the new industries relied more on training and skill, workers were less easily discarded and replaced. The leverage of these workers increased. The new industries were tied much less than previously to products which had to be sourced from the rural estates—hydroelectricity trumped coal, steel trumped timber—so the relative power of the aristocracy was further weakened by the changes. Everybody but the aristocracy grew stronger.

During the twenty years or so of recession, harsh conditions provided fertile ground for the growth of class consciousness and rebellious ideas among the working class. Their members had starved as the old industries failed. In many cases, their desperation had hardened into anger. As the new industries began generating wealth, worker demands for better wages, safer working conditions, and an expanded franchise were louder and more organized. Each gain they achieved justified the power they felt.

5. The Inevitability of Empire

Even though the aristocratic class felt threatened by both the working class and the magnates, they were not totally powerless. Some of the raw materials on their land still fed significant benefits back to them even though the share of the economy they controlled was dwindling fast. The major sources of their strength, however, lay in the remnants of their past dominance. They had claimed to be the protectors of society, so they still had active control of the military. The legal system and parliament had been their invention to help them harvest wealth and they still had preferred access to both—the House of Lords gave them a veto over any formal change, the House of Commons was dominated through an exaggerated weight given to rural ridings, judges were aristocratic appointments, the police were sworn to enforce laws favourable to them. The aristocracy could move in these areas to ward off threats to their position.

They doubled down on their traditional position as embodiment and defender of the nation by stepping up the rites, the pomp and the boasts which proclaimed their indispensability. They spent lavishly on visible symbols such as stately homes, gaudy uniforms, balls and parades. However, this would all ring hollow if they were not also seen to be the defender of the nation. And a defender needs an enemy to make their story believable. The elites had to increase the aggressive international swagger of their nation. This led to moves in parliament to re-arm and build up the military. Every state around them noted this activity and followed suit, wrapping belligerence in their respective flags, blessing it from their respective pulpits and dressing it up in their own flavour of pageantry. Once warfare was seen as a real possibility, they could use the military and the police to crush any domestic dissent in the name of civic solidarity.

Of course, the ensuing arms race fed on itself. Threats at the borders, initially an exaggeration to bolster the authority of a fading class, turned into reality. As historian Arno Mayer has written about the origins of the First World War: "War ceased to become the continuation of diplomacy to become the extension of politics, Europe's governors becoming ever more prone to resort to foreign conflict to further domestic objectives."[3]

The industrial magnates enthusiastically supported this course because rearmament produced lucrative contracts for their new industries. The middle class also leapt on board as they too found profits in this flurry of activity. The industrial workers had less reasons to follow, but the jingoism, the pomp, the pageantry, the pulpit, the police, the military, the increasing number of jobs and a centuries-old attitude of obedience swept them along as well. For the last part of the 1890s and the first decade of the 1900s, the fact that the real distribution of power was out of step with the structure of privilege drove a growing militarization on the whole European continent.

England was not alone in invoking war to shore up her fading ruling class. The Romanovs in Russia, the Hapsburgs in Austria-Hungary, the Hohenzollerns in Germany and the anti-republican forces in France were all staving off the future with the same strategy. The situation was more volatile on the continent because those countries, being less wealthy, had a smaller pie to split. They lacked England's empire as a reserve of wealth. Their chosen enemy was just across an imaginary line on a map instead of across a channel. And, because of a century of slower growth, continental workers generally had a more militant working class. By 1914, the five empires of Europe were all armed to the teeth while they hurled insults at each other. The two supposedly defensive alliances which arose from this fearful situation, Austria-Hungary with Germany and Russia with France and England, were collectively capable of immediately putting eight million armed men into the field. It was ironic for the aristocratic class that each escalation in armaments only added strength to the new industries, increasing the power and numbers of both the magnates and the working class. Europe was a powder keg. The shot in Sarajevo lit the fuse. As Mayer observed:

> ... the governors of the major powers, all but a few of them thoroughly nobilitarian, marched over the precipice of war with their eyes wide open, with calculating heads and exempt from mass pressure ... from which they hoped against hope to draw benefits for themselves."[4]

England entered the fray because the aristocracy hoped that this war, like others in their past, would reaffirm their claim to power. All eyes were turned towards them as they set out to be the defenders of the realm.

However, this time things were very different. New weapons based on the new industries of steel and industrial chemicals had transformed the battlefield. The old tactics were trotted out again by old aristocratic generals but they only served to plunge the battlefield into a slaughterhouse with no end. Each combatant sacrificed huge numbers of lives and squandered massive amounts of financial resources. The empires recalled their armies from the colonies, handing subject peoples an opportunity to escape imperial control. The revenues, resources and markets that the colonies had provided disappeared. The cost of the war and the lack of revenue from the colonies ruined the aristocracy while the new factories were working flat out producing armaments, making the workers essential while enriching the magnates.

The phenomenal loss of lives with none of the promised glory or territory caused any residual faith that people may have had in the aristocracy to collapse—just the opposite of what had been intended. Seen by everyone to be helpless as defenders and incompetent as leaders, the validity leached out of their claims to be the soul of the nation. Pomp segues into farce when rituals become disconnected from reality. The aristocratic class had proven decisively and publicly that they were no more than the remnants of a caste of feudal fighters who had become irrelevant in the modern world. Being the protector of the nation was at the heart of their legitimacy. Without this, they were finished. In a world of mud, tanks and machine guns, an aristocratic regiment showing up with beautiful horses and polished sabres was an insult to the men being ordered over the top. Their inability to come to terms with new weaponry while refusing to give up inherited positions of command prolonged the war, causing millions of extra deaths. More and more as the war dragged on, they had to be sidelined in order to bring the war to an end.

The old regimes in each of the participating countries lost their bet that the conflagration would reaffirm their positions. The future they had been attempting to forestall arrived with a vengeance. By the end of the war, the Romanovs in Russia were dead and gone, replaced by Bolshevik committees. The Hapsburgs, after controlling middle Europe for half a millennium with a completeness that made it impossible for people to conceive of the region without them, disappeared like snow on water, replaced by a handful of nationalistic states. Germany was shaped at the peace conference into a tarnished, impoverished, toothless republic. In England and France, control of the political process moved decisively from the hands of the aristocracy to a much more widely enfranchised and representative parliament. Though the monarchy and the aristocracy in England survived with their lives, some of their lands, their titles and their silly uniforms, everyone now knew that they were museum pieces—relics of a bygone age. The Great War was the dying gasp of Europe's feudal aristocracy.

It was inevitable that England lose her position as the eminent power in the world when the second industrial revolution produced the steel warship powered by coal. Her dominant navy became instantly obsolete. Any country with resources could use this new technology to challenge for supremacy of the seas. Her empire was in tatters. Without their resources, she was no longer rich, no longer powerful. She was no longer important. To follow the forces shaping the world after World War I, we will need a different lens. And at the end of the Great War, there was an obvious choice, a young country bursting with strength, lying across the Atlantic.

In each of the destroyed countries of Europe, two groups—the industrial working class and the industrial magnates—were pushing to fill the vacuum that had appeared at the apex of the power structure. It was not immediately obvious which would be dominant. The relationship between them needed more time to sort itself out. Meanwhile America, largely unbloodied in the conflict, was flexing her muscles. She also had workers and industrialists but the relationship between them was very different than was the case in Europe. She was strong, confident and dynamic, thrusting vigorously onto the world stage. We will now examine her story to see how the evolving relationship between labour and magnates took us from the end of the First World War to the present.

With society everywhere fractured into two major antagonistic classes, the workers and the magnates, and two minor ones, the educated middle class and the aristocratic remnant, there could be no overarching allegiance to the whole of society. Each group nurtured cooperation within itself, offering esteem to those creations that benefitted other members of their class. But actions exhibiting care for the whole of society would benefit "others" and this was anathema to all. Common resources became assets to be claimed. The commons became a place that could be despoiled without compunction. The lack of unity exemplified in this world of competing power bases made cooperation beyond the limitations of one specific class impossible. We are reaping the tragic consequences of this historical split.

Chapter Six

The American Empire

The eastern seaboard of America was originally colonized by people willing to suffer to escape aristocratic control. The Mayflower carried religious non-conformists seeking the right to worship as they chose, but the colonies also became an outlet for a range of discontents and political dissenters.

The British nobility tried to fasten a traditional form of feudal control over this imperial outpost, but they failed. When the Duke of Baltimore was given what has since become the state of Maryland as a fiefdom, he found that he had land but no peasants to work it. The natives had drifted westward rather than stick around to be enslaved. The colonists already present had escaped dominance once and valued their status as free men. And peasants would not uproot themselves from the old country unless a significantly better life was on offer than the one they already had at home. Even if he did convince some intrepid settler to take this step, there was always the temptation of arable land which could be claimed beyond the western borders of the noble writ. It proved impossible to either produce an estate system based on enslaved natives, as had happened in India, Africa or the Caribbean, or to reproduce the feudal system with transplanted colonists from England.

If a chance to own their own land was on offer, however, there were definitely volunteers, people prepared to risk a chancy ocean crossing and an uncertain future farming in an unknown land. Most of the eastern seaboard was well suited to agriculture and soon there was enough harvest to ensure survival. Such farms, generally worked by the owner himself, were necessarily small. Siting these farms in groups for mutual aid produced farming communities, groups of fields and a few buildings carved out of a wilderness of woodland. The lure of freedom was strong and it continued to attract a steady flow of immigrants. These isolated settlements grew. More of them appeared. The frontier pushed westward. There were no peasants and no aristocrats, just farmers.

The natives having been pushed west, there was no conquered people for the colonists to administer and exploit. This America of small farms had little need of a military class to police an enslaved population as had been necessary in India, Africa, the Caribbean or South America. This lack of a significant resident military presence was crucial to the developing character of the colony. The typical American in the early 1700s, especially throughout the north, was a free farmer who owned his own land, answering to no one and commanding no one. This had not occurred since the days of classical Athens.

In the south, things were starting to develop differently. The land and climate favoured the growing of cotton, a product that was becoming valuable to the new mills in Britain at this time. Rich colonists could claim large parcels of land and for labour could tap into the slave trade that was currently servicing the Caribbean sugar plantations. This in turn made a permanent military presence necessary because slaves need to be intimidated and controlled. Slavery, large plantations with subjugated workers, and the consequent need for a resident military force pushed this area towards the recrystallization of a militarized aristocratic class. The different paths of development between north and south portended trouble.

In the north, however, widespread equality became the norm. Land was generally owned by the people who worked it and ownership of land was within the reach of most colonists. Each small community governed itself; everyone in these communities had a say in that governance. And because it was possible to pull up stakes and "go west" to a spot where they would be answerable to no one, every member of each community was a member by choice.

Leaders in the community were people who had demonstrated that they were worth listening to. Anyone, no matter how circumscribed their station in the old country, could become a pillar of one of these new communities if their personal qualities earned them esteem from their neighbours. There was one class—no slaves below and no elites above. Having to rely on their community made them a tight group: they had to listen to their urge to be the Cooperator. The nature of the land, the self-selection for a non-conformist colonist, and the steady arrival of new colonists which slowly moved the frontier west, made equality and freedom lived experiences. As had happened in Athens and the European city-states, broad equality and a life based on cooperation called forth great things.

The chance to live free was such a powerful attraction that the colony grew quickly. Because the Atlantic Ocean lay between them and the English parliament, direct control by Britain was weak and was centred almost exclusively in the seaboard cities. As long as the cotton, foodstuffs, timber and furs were available to British traders, and as long as the colonists continued to be a reliable market for English manufactured goods, the people were largely left on their own. Organization was based on the towns that anchored each agricultural area. Though the people nominally owed their allegiance to the king, in practice they solved their own problems, making their decisions in open town meetings. All adults were expected to contribute to these decisions and to their implementation. Broad equality in Athens had called a similar political structure into being.

In each community, a few individuals prospered and began to accumulate wealth. When they looked around for investment opportunities, they found themselves obstructed at every turn by British laws and taxes put in place to give precedence to British merchants. For example, the Navigation Acts gave British traders control over all trade into or out of the colony. New Englanders wanted to share in the rum, sugar and molasses importation from the Indies, the transporting of slaves from Africa, the whale and seal hunts, and trade along the coast of the Americas and into the Pacific, but these ambitions were being thwarted. Discriminatory taxes inhibited the development of local manufacturing, attempting to limit the colony to being a provider of raw materials and a purchaser of finished goods. These grievances accumulated. Meanwhile, the colonists continued to grow in both numbers and wealth. A significant number of men were familiar with guns because hunting was a regular part of their lives. It was no longer obvious that Britain could enforce the restrictive laws that operated to its benefit. The colonists rebelled.

The principles for which the colonists were prepared to fight reflected the love of freedom that had initially drawn them across the ocean. These principles, as asserted in the Declaration of Independence, framed the American Revolution as a challenge to the whole edifice of feudalism.

6. The American Empire

All men are created equal. All men have been endowed by their creator with an inalienable right to life, liberty and the pursuit of happiness. The purpose of a government is to safeguard these rights for the individual. Governments derive their just powers from the consent of the governed. There is to be no aristocracy. There is to be no church of state. Freedom of speech is guaranteed, especially the right to criticize those in government. There is to be the opportunity for universal education. In practice, because of the growing reliance on slaves in the cotton and tobacco estates, these lofty goals were applied to white men only. Nevertheless, these ideas mark a breathtaking leap from the norm in 1776 where in Europe people were being forced to work a master's land in various forms of servitude. Elsewhere around the world where colonial powers had touched down, people were being hunted as vermin or captured and sold into slavery.

The framers of the Declaration of Independence—the leaders of this new society who were starting to accumulate wealth, many of them from the northern towns but also some owners of estates worked by enslaved people in the south—might have chosen to affirm a life of privilege for their small group, embraced the plantation model, made common cause with British troops and promoted themselves as a new aristocracy. But they did not. Most of them had only recently earned status amongst their fellows; they were still embedded in their communities and their identification with their communities was strong. The desire for equality and freedom that had drawn them across the ocean still resonated strongly with every one of them. They had read their history and had been dazzled by the examples of Athens and Florence and Venice. They chose to make their declaration a document which celebrated equality.

The southerners on their plantations wanted a relaxation of English trade restrictions and imperial taxation so they could sell their cotton more profitably. The town leaders from the north wanted to be able to invest, build mills, control their own imports and exports, and grow their wealth in ways that were currently frustrated by British law. The British right to rule over them was challenged. Britain accepted the challenge, fought for the right to order society in their favour, and lost. The colonists became truly free.

With the removal of the restrictive British laws and taxes, the pent-up potential in the north allowed it to leap forward economically. For example, in 1790, US owned ships already handled 58 percent of its foreign trade while by 1807, this proportion had risen to 92 percent.[1] Manufacturing also increased but, lacking a large domestic market, unable to challenge Britain for world markets, and with investors holding only modest fortunes, industrialization took the form of small mills producing mainly for local areas. However, the existence of these mills increased the variety of local jobs and generated profits so the number, the wealth and the population of these local areas increased steadily, continuing to draw in colonists. The population of America grew steadily throughout the 1800s. These new people expanded the market for her manufacturers.

But government remained overwhelmingly local. As writer Paul Goodman described:

> When the American revolution removed the British authority at the top, society remained mainly organized as a network of highly structured face to face communities and associations, and these were fairly autonomous ... Democratic or hierarchic, the groups were small, and, on most matters, people were in

frequent personal contact with those who initiated and decided.[2]

Town meetings and local associations provided all the political organization that was necessary.

The south got rid of the onerous taxes on cotton, but independence did not unleash growth to the same extent as it did in the north. Their economic success led to ever larger estates and a cotton aristocracy who imported their luxuries from elsewhere. Immigrants to the south didn't add to the market as they were mostly enslaved people. The extent of the large plantations kept towns small and dispersed, further stifling the growth of local markets for manufactures. Consistent with the coalescing of a nascent aristocratic class, the wealth the plantations generated went into display rather than economic growth. These differing economic trajectories meant that the north soon dwarfed the south in both numbers and wealth. Their growing differences strained the cohesion of the union with the issue of slavery coming to symbolize an unbridgeable gap between them. For those in the south, it was seen as the essential basis of their economic existence. Northerners could not compromise on this issue without denying the desire for freedom that had initially inspired them to cross the ocean into the unknown. This appreciation for freedom still informed their daily lives through the workings of their communities. The ensuing civil war accomplished in America what the First World War did for Europe. It decisively ended the aristocratic power in the south while giving a mighty push to the rising industrial forces in the north.

The Civil War provided good business for manufacturers of everything from weapons to uniforms. At its end, the government was in debt to merchants and bankers in the north and when these debts were paid off by general levy, significant capital suddenly became concentrated in the hands of investors. The timing was magnificent. At the first stirrings of that economic leap based on steel, oil, electricity and industrial chemicals, America had a class of merchants with money to invest and no aristocratic remnant to hamper their activity. For many of these industries such as steel for rail expansion and chemicals for use in munitions, the Civil War had provided the market to help American producers grow these industries at their very first stirrings.

The Civil War also greatly increased the power of the federal government over local institutions. War requires a controlling and organizing power. The independence of the town meeting was eroded as distant governments took on many of the decisions and prerogatives that had previously been theirs while demanding taxes in return. Also, the gap between the wealthy and the poor was increasing as the manufacturers thrived, the towns and the middle class grew, and the regions became less dominated by agriculture. The general sense of equality was weakening. Class interests were reasserting themselves. As in Athens and the medieval city-states, however, the change happened slowly and the structures that embodied every citizen's identity with the town continued to persist for quite a while as did the mythic memory of equality as a central fact.

America may have gained a head start with their Civil War but the European powers were the big boys on the block—they still had dominant navies, empires to exploit and reserves of capital that had been accumulating for centuries. They were hindered now because the struggles between them directed their focus on rearmament and this gave the American industrialists an opportunity to become leaders in these new industries if they could grow quickly. They had capital from war profits but they needed both larger markets than America was providing and a ready supply of cheap labour to build the railroads, to construct and man the new factories and to descend into the mines. Citizens whose forebears had crossed an ocean in search of freedom and then had fought both a revolutionary and a civil war to defend it would not easily submit to the exploitative conditions on offer. Finding a cheap and pliable workforce became their first pressing problem.

The opportunity to homestead on the western borders became more difficult after 1865 because America was finally running out of frontier. People could no longer just walk away from bad situations and pre-empt land. They had to make do with the conditions they faced where they were. In Europe, the depression of 1873-1896 was devastating. Many workers lost their jobs; families slipped into poverty and faced starvation. They had all heard that in America, a land without aristocrats, a man's future was limited only by his willingness to work hard. The Emma Lazarus poem on the Statue of Liberty soon requested, reflecting America's open door to that pool of unskilled labour: "Give me your tired, your poor, / Your huddled masses yearning to breathe free, / The wretched refuse of your teeming shore. / Send these, the homeless, tempest-tost to me, ..."

The huddled masses came. They were huddled again on Ellis Island, and then they continued to be huddled in the slums of the growing industrial cities, needing work, any work, under any conditions, to keep themselves alive. They were a few generations too late to seize the myth that had inspired them. By the time they realized this, it was impossible to go back. The yearning to breathe free would have to be focused on the next generation.

The American industrialists rode the wave of the second industrial revolution. Having items under production early allowed them to enter international markets once they serviced their domestic market. The westward expansion of the railroads allowed them to exploit sources of oil, gas, coal and other minerals on land which they could control cheaply. And there was a flood of desperate immigrants to do harsh, dangerous, unskilled jobs for a pittance: between 1865 and 1914, the population of America grew from thirty-three to ninety-one million people, with most of these newcomers bringing little education, a strong back and a necessity to work. Driven by ample capital, an excess of labour, and the exploding markets for these entirely new products, with no organized labour presence or aristocratic remnant to counter their activities, there were no brakes on the growth of industrial America. Initial investment was quickly repaid and reinvested. The robber barons, as the tycoons of the post-Civil War era have come to be known, grew so powerful that they lived beyond the control of either the marketplace or the government. As historian C. Wright Mills described them:

> The robber barons ... exploited natural resources, waged economic war among themselves, entered into combinations, made private capital out of the public domain and used any and every means to achieve ends. They made agreements with railroads for rebates; they purchased newspapers and bought editors; they killed off competing and independent businesses, and employed lawyers of skill and statesmen of repute to sustain their rights and secure their privileges ... the general facts are clear; the very rich used existing laws, they have circumvented

and violated existing laws, and they have had laws created and enforced for their direct benefit.³

It was a grand time to be a magnate but a brutal period for the industrial worker.

The population and the amount of industrial activity were expanding so rapidly that large numbers of new service, management and professional positions were also being created. With immigration aimed at securing unskilled labour, these middle class positions were mostly filled from within the country. In this pool of opportunities, the workers saw a way out of the slums for their children. The progression was so common as to be visible. Each new wave of immigrants took over the dangerous, strenuous and poorly paid jobs. They moved their families into the slums which were being vacated by the previous wave of immigrants who, in one generation, had advanced in education and accent so they could now attain a middle class position and move to a better neighbourhood. The dream of having horizons limited only by one's willingness to work hard turned out to be true after all. It had just turned out to be a multi-generational process.

This constantly visible possibility of a better life for their children convinced most immigrants to endure the harsh conditions and focus their hopes on the next generation. Because of this, no militant working class organization took root in America. In Europe, slower rates of growth over a longer period had allowed the middle class to become largely self-perpetuating, so there were few professional positions available to working class children. There was no visible route to a better life for the children of European workers other than dramatic systemic change. In America, the workers kept their heads down, their noses to the grindstone and their shoulders to the wheel, persevering so that their children could move into the middle class. The robber barons reigned unopposed.

This short period of untrammeled growth, which lasted from about 1870 until the First World War—a scant forty-five years—fed the strong American belief that "getting ahead" is the goal and that the measure of success in life is that your children fare better than yourself. Many families do have just such a history from this period; these family stories strongly reinforce the myth. The fact that this possibility is available to a broad segment of society only during a period of rapid growth under a transforming industrial landscape has been lost from the story, dooming many today to feelings that they are not measuring up.

As the 1900s began, the markets for the new industries started to saturate. The European magnates entered the new sectors with massive amounts of investment and they could quickly dominate in the protected markets of their empires. Domestic markets were mostly serviced. Railroad building was easing off—the most profitable lines were in place. The automobile was still a curiosity, not yet an industry. So, like England before her, America had to look towards empire to market her goods and to find greater investment opportunities. As the British historian J. A. Hobson summed up:

> The adventurous enthusiasm of President Theodore Roosevelt and his 'manifest destiny' and 'mission of civilization' party must not deceive us. It was Messrs. Rockefeller, Pierpont Morgan and their associates who needed imperialism and who fastened it upon the shoulders of the great Republic ... They needed imperialism because they desired to use the public resources of their country to find employment for their capital which would otherwise be superfluous.[4]

The urge to dominate drove these men and they controlled the levers of government. America, the great republic whose founding documents were a paean to freedom and equality, moved aggressively as a bully into the international arena.

A flavour of the period can be gleaned from a 1935 article published in the magazine *Common Sense* by a disgruntled retired marine corps officer, Major General Smedley D. Butler, which reads:

> Thus, I helped make Mexico, and especially Tampico, safe for American oil interests in 1914. I helped make Haiti and Cuba a decent place for the National City bank boys to collect revenues in. I helped purify Nicaragua for the international banking house of Brown Brothers in 1902-12. I brought light to the Dominican Republic for the American sugar interests in 1916 ... In China, in 1927, I helped see to it that Standard oil went its way unmolested.[5]

America could continue to grow economically by expanding her sphere of influence into Latin America and the Pacific. The growth of this empire in the early twentieth century provided some respite from the dilemma of accumulating capital. The European powers still firmly controlled most of the world. America had to make do with a few areas along the fringes that they could enter without risking a major conflict.

Once again, the timing of American activity was extremely fortuitous. Just as she was moving outwards into the world, the European powers were retrenching, focusing on rearmament and the belligerence that would soon explode into World War I. American robber barons, with neither an aristocratic remnant to reassure nor a significant working class to subdue, were free from the pressures that had set the European powers on the road to war. During the war itself, America could remain aloof, her industrialists marketing military supplies to everyone, expanding their industries and steadily increasing in wealth. Meanwhile, the European powers were tearing each other apart, sacrificing the best part of a generation and most of their bullion. At war's end, their empires were left weakened and largely undefended. By 1917, when America finally sent troops, no combatant had much appetite left for trench warfare and the fresh American troops could break the stalemate in favour of the allies. But having been the power that forced a resolution, America could strongly influence the terms of the peace. As historian A. J. P. Taylor noted, "American policy was never more active and never more effective in regard to Europe than in the 1920s. Reparations were settled; stable finances were restored; Europe was pacified; all mainly due to the United States."[6]

Germany owed reparations to France and Britain. France owed debts for war materials to Britain and America. Britain owed similar war debts to America. The money all flowed across the ocean to America, to her bankers, to her producers of armaments and to the oil, mining, steel and chemical industries that had leapt in size thanks to the massive demand for munitions.

With the end of the war, new investment opportunities for American bankers and industrialists appeared like mushrooms after a rain. They gained access to the orphaned colonies of the old powers where they could take over the existing command structure and immediately start to extract resources and exploit new markets. There were also vast opportunities in rebuilding the devastated European continent itself. The American steel mills, mines and factories expanded as quickly as capital and labour could be made available. The capital was flowing in from war debts but the old problem of a lack of cheap labour raised its head once again. Immigrants were no longer crossing the Atlantic: war deaths had decimated the European work force and reconstruction was providing jobs for those who had survived. Also, the American myth had been tarnished by their recent history of labour exploitation. The huddled masses stayed home. A new underclass to perform the dirty jobs was essential and, for the first time, it looked like it might have to come from America itself. Fortunately for the magnates, the solution was at hand.

New agricultural machinery such as tractors and automated cotton pickers were making many southern plantation workers redundant through the twenties, thirties and forties. These farm labourers and sharecroppers already had such precarious lives that the possibility of industrial work in the north or on the west coast, badly paid and dangerous though it promised to be, could still look attractive compared to their present prospects. The push of desperation and the pull of an available job caused millions of people to flow from the farms, many of them in the south, into the industrial cities in a movement that has since been labelled "the Great Migration." These rural newcomers replaced the last wave of pre-war immigrants whose children were now rising into middle class jobs and neighbourhoods. Many of these farm workers were black, the descendants of slaves.

The rapid industrial growth after the end of the First World War provided jobs for these displaced farm labourers. Work on an assembly line had more security, was better paid and had better prospects for the worker and his children than could be found sharecropping or working seasonally as a farm labourer. The move to the city, difficult as it was, generally improved their lives.

These displaced agricultural labourers were the source for post-war unskilled labour. Because many of these workers were black, the white men making policy wanted to keep the gulf between the middle class and the workers dauntingly wide. They restricted tools such as education that would make class mobility easier and instead started to favour educated white immigrants who could slot directly into positions in the expanding middle class. The traditional way out of the slums on the social escalator was taken away. The new underclass, unlike those generations that had preceded it, was expected to remain permanently in that position. The rapid increase in the numbers of these industrial workers, both white and black, now speaking the same language and having little expectation of seeing their children leave this underclass, finally produced a movement that wanted to use their massed numbers to advance their demands. They had successes. Real wages rose slowly but steadily throughout this period.

Middle-class opportunities that were not filled by educated white immigrants went to the children of white workers because they could assimilate more readily into middle class workplaces and neighbourhoods. Many American cities became increasingly split between the underclass of black, poorly educated, unskilled workers in poor neighbourhoods and white owners and middle class professionals in their leafy white suburbs. The influx of unskilled labour from the farms made the boom times that followed the First World War possible. But the way in which it occurred ensured that race would continue to be a major source of discord in American life.

The unopposed industrial growth that defined the period from 1860 to 1920 had caused production to expand many, many times. The profits that the magnates needed to reinvest were greater with each successive year. The factory world was growing explosively. But, as was the case a century earlier in England when capital and labour were abundant, this drove a need for markets to expand commensurately. By the end of the twenties, the markets for American industrial production were saturating. There were difficulties selling everything that could be produced. Good investment opportunities were becoming scarce. More capital than sound investment opportunities drove down returns on all investment, old as well as new. The depression of the 1930s had the same root cause as the great deflation of 1873 to 1896 in Europe—a lack of investment opportunities in the face of accumulating capital. By 1929, this competition to make capital useful was undercutting profits that had once seemed secure. Weaker, more leveraged, more speculative businesses failed and laid off their workers. Fear of unemployment caused people to stop buying and markets restricted even further, causing more businesses to fail and more people to lose their jobs.

The collapse occurred virtually instantaneously because complacency had led everyone to devalue the risk of highly leveraged investments. After sixty years of unchecked expansion, everyone had come to assume that growth was guaranteed. Meeting desires with a down payment and a promise to pay the rest in the future was seen as the same as having cash in hand. Cars, houses, furniture, stocks, businesses—everything rested on credit. When a few businesses failed because markets were saturating, lenders started calling in loans, but there was no cash anywhere. A borrower or lender in trouble toppled a chain of others who owed money they did not have. This made failures roll through the economy, but the underlying root cause was saturating markets.

In the economic world, one man's disaster is another man's opportunity. Companies that had already attained a monopoly, either singly or in a colluding group, could still guarantee their profits. They could scoop up assets of failing businesses. The depression of the 1930s, as did the First World War and the Civil War before it, concentrated ownership in the industrial sector into fewer hands with larger holdings. A few monster firms were starting to dominate each sector.

We are now ready to pick up the drama we set aside at the end of the last chapter and examine the relationship between the magnates and the industrial workers as they each jockeyed for the dominant position once held by the collapsing aristocracy. Both emerging classes had seen their power validated by contributing essential aspects to the defence of their society. However, it was still not clear whether it had been the industrial workers taking to the field in their millions or the magnates conceiving and delivering the essential weaponry that had been the essential factor.

The magnates' power rested on their ability to direct the wealth produced by the economy. They had focused this on producing the tanks, airplanes, machine guns, artillery shells, bullets and bombs without which their army would have been helpless. Controlling pay cheques gave them a point of direct control over the workers while controlling the dissemination of information through ownership of the media allowed them to shape the narrative by which many people ordered their lives. Their ability to protect the weakened aristocrats from the masses gave them access to traditional aristocratic power centres such as parliament, the military and the police.

The workers' power lay in their concentrated numbers, especially in the manufacturing cities. The prospect of them mobilizing as a mob and overwhelming the less numerous magnates, aristocratic remnant and middle class was always hovering in the air. The war had proven that military success relied on numbers, victories coming only after wave after wave of soldiers had perished. Working men in infantry kit had been the ones who had killed and died. Industrial innovation had made this soldier as potent as possible by refining the repeating rifle and the grenade and those technologies, in turn, multiplied the effectiveness of the mob. Just as had happened with the invention of the phalanx, the invention of repeating rifles and small bombs moved power away from the few with wealth and towards the many with cheap personal weapons.

The struggle was complicated by the fact that each saw their need for the other. The industrial workers would starve without the factory system providing the innovation, the organization, the pay cheques and the consumer goods on which they relied. Conversely, the factories would be useless to the magnates without workers to man the assembly lines. It was a dilemma for both.

In America, the workers had arrived from many lands with their own languages and traditions and this initially isolated them from each other. The possibility of the next generation rising into the middle class had then induced most to accept the harsh status quo for the sake of their children. But when the hope that their children would ascend to the middle class dimmed, and when the Depression caused many to lose their jobs and face hunger, American workers finally realized the same bleak truths that had defined life for their European counterparts for more than a century. Many American workers had also been soldiers, knew the power of a multitude with guns, and were reluctant to accept that this hardship was inevitable. In the thirties, American workers began to explore the power inherent in their numbers. They discussed challenging the descendants of the robber barons for control of the economy as a real possibility.

The American magnates, however, were in an extremely strong position. They had been expanding their wealth and power unopposed by workers or government since the end of the Civil War. They controlled the media, the military, and the political process and they had the unquestioned support of a substantial middle class which they reinforced by reframing class demands in racist and xenophobic terms. They were up to any immediate threats from these fledgling worker organizations.

In Russia, the small amount of industrial production that existed was accomplished under imperial auspices so there was no class of non-aristocratic magnates to speak of. There was a small aristocracy, a mass of dispersed agricultural peasants who produced the wealth that fuelled aristocratic excess, and a fairly small group of ill-treated workers in imperial factories, concentrated in a few of the major cities. When the aristocracy was critically weakened by the war, the workers focused their violence directly onto the dying aristocracy. The aristocracy collapsed and a small cadre of industrial workers seized control. The machinery of government they inherited had been built to control and exploit the peasantry. As in Sparta, they just adapted the existing system to continue removing wealth from the peasantry, first in the name of short-term capital accumulation, but soon in organized permanent exploitation. A new elite quickly emerged from the leaders of the working class. One aristocracy replaced another.

In Europe, however, things were not as clear cut as they were in either America or Russia because the two forces were more evenly matched. The working class was stronger and better organized than it was in America, with a tradition of struggle reaching back to the beginning of the Industrial Revolution. They had a history of popular uprisings since the French Revolution, a pantheon of martyrs and a body of analysis that defined their position and their goals, all of which hardened their commitment to engage in a fight for control. Because the war had been fought in their backyard, millions of workers had become familiar with rifles and everyone, soldier and civilian alike, had experienced the effectiveness of masses of armed individuals in effecting change. The industrial workers were a formidable force.

Because of this real threat, firm and reliable control over an obedient military was the essential element in any effort by the magnates to establish themselves as a new elite. Their dilemma was that this army would have to draw its members from the very working class that the industrialists needed to control. How this could be accomplished was key to the struggle.

After the war, the European economies were devastated. Many factories had been destroyed. The resources from the colonies had been cut off. A focus on production for war had distorted industrial output. Paying down war debts removed so much capital that there was none left for reinvestment. This removal of capital also made inflation and unemployment inevitable, pricing many goods out of the reach of many. Every economy struggled. Unemployment, poverty and discontent were widespread. Through the twenties, this poverty persisted and anger simmered. The 1926 general strike in Britain and the Weimar hyperinflation and instability of the early twenties in Germany were visible signs of this discontent.

Making this situation even more volatile, each country had millions of demobbed soldiers, many of whom had been psychologically wounded in the trenches. They called it shell shock; now it is known as PTSD. The hell of the trenches with horrific deaths multiplying around them year after bloody year, each man waiting for their turn to be ordered over the top into the killing zone, defined a living nightmare that left no man unscarred. When the war ended in late 1918, the survivors were thrust back into a world in which they no longer fit. Even if they could manage to hold a job, the nightmares, the neuroses and the compulsions simmered just beneath the facade of daily life. The First World War wasn't over in 1918—it pushed its long tainting tentacles into the twenties with the irrationality of this damaged generation. This was less true in America as US soldiers spent less time in the trenches and the numbers of damaged soldiers represented a much smaller part of the total society. But in Europe, shell-shocked citizens were a large part of the new reality.

There was a long history of private mercenary armies called Freikorps in central Europe. Because this region had been made up of many small principalities who were frequently at war with each other, these mercenary armies, owing no allegiances except to their leader who would seek out the highest bidder, had existed since the early 1700s. Being in one such group was an attractive alternative for demobilized, damaged soldiers who could feel comfortable nowhere else. The camaraderie of a group that shared their private hell assuaged their feelings of isolation; the promise of group violence released their suppressed anger. In Germany, a number of these organizations formed around psychopathic former officers as the soldiers limped home. Loyalty to these leaders was neither logical, ideological or political but, based as it was on a psychological need, it was strong, unshakable and unquestionable.

In every country in Europe, the industrial working class split between the left and the right, the proffered communist and fascist solutions to their problems. The left, growing out of the traditional labour movements, based their solution on replacing the magnates and gaining direct control of the factories and the political process themselves. They had the examples of having already forced a steady expansion of legislated rights as a template for an incremental march to utopia on the one hand, and of a group of workers seizing sudden control in Russia as a harbinger of immediate change on the other. Their dream, their analysis, their demands and their organizations all stressed equality and solidarity. This appeal to the Cooperator felt intrinsically right to them. Many of the returning soldiers, however, had a much more bleak, individualistic, nihilistic, violent view of the future. They sought comfort in an unequal hierarchical world where, as it had been in the army, a few individuals would make the decisions and tell them what to do, relieving them of thought or responsibility. They were only comfortable with this fascist view of the future, a view shaped by the Dominator.

The growth of the right in the working class occurred markedly in Germany but fascist organizations also prospered in Britain, Italy, France and elsewhere. Adolf Hitler was one of these damaged returning soldiers. He had been both wounded and decorated as a message runner in the trenches. Before the war, he had failed in his dream to become an architect and for that he blamed an aristocratic Prussian state which shut out people with talent but little money and few connections. He had found a home and a purpose within the army.

Though Germany was being overwhelmed on the battlefield, the actual ending of the war had been precipitated by a revolution by sailors in Germany's North Sea fleet in the fall of 1918. This unrest soon spread to the cities where it was taken up by the communist and social democratic unions. The government fell. The Kaiser fled and a newly installed left-leaning civilian government had sued for peace. The army, having not been allowed to fight to the last man, clung to the fiction that they could still have prevailed had they not been "stabbed in the back" by left wing politicians. That version of history, painting the soldiers as victims rather than losers, inflamed these damaged returning soldiers against the left. The masses were split. One group wanted a structure that encouraged and empowered the Cooperator; the other fought for a world that liberated the Dominator. Two very different futures hung in the balance.

The harms that the war had inflicted on Hitler presented as paranoia, uncontrollable anger and megalomania. These attitudes may already have existed before the war, results of the failures of his dreams, or because his father beat him or that his mother had died when he was young, but the war had forged these affronts into manias. He uncovered a skill at stream-of-consciousness, hate-filled flights of oratory, and this skill helped him take over a nationalistic, racist trade union with a handful of members and turn it into a political party. Renamed the national socialist workers party, it grew because his performances attracted like-minded, angry, bitter ex-soldiers, including the leaders of a few Freikorps battalions whose members, biddable sources of violence, became the nucleus of the Brownshirts. The essence of his message was nationalism—German exceptionalism, mythic history, the glorious future owed to this superior race—and scapegoating all those who, in his view, had stabbed the army in the back.

A communist rising in 1919, inspired by the horrendous conditions at home and the example of the events in Russia, attempted to unseat the social democrat government, but it was put down by combined efforts of the installed Weimar social democrat government, the post-war professional army, the magnates' wealth and a frightened middle class. German politics was now clearly split into

6. The American Empire

three groups—this governing alliance, the defeated communist believers with their control of many of the trade unions, and the violent fascist parties. The governing alliance was dominated by the magnates as their wealth gave them control of the middle class, the media and the courts, and influence among the officer corps of the army, law enforcement and the factories as employers. With the lesson of Russia still fresh, the communists were seen as the greater working class threat, so the magnates favoured the fascists, enforcing laws against left wing violence while turning a blind eye to brownshirt excesses. Communist challengers for power were shot while Hitler got a year in a comfortable prison to write his memoirs after the failed Munich Beer Hall Putsch. Being allied with such violent fellow travellers served the magnates' purposes. They believed they could control the fascists when that need arose.

When the unsettled decade of the 1920s morphed into the depression of the 1930s, however, things soured for everyone—many businesses failed, inflation eroded all wages and unemployment rose dramatically. Discontent escalated. Both the communists and the fascists had structures available to focus that discontent.

The German economy was extremely fragile because reparations had drained their financial reserves and triggered hyperinflation, driving much of the middle class into destitution. Business failures and unemployment escalated while a bankrupted government could offer no help. The magnates were becoming weaker while both the communists and the fascists found these conditions conducive to securing new adherents and making violent solutions seem more palatable. The magnates had to follow their acolytes and embrace nationalism in order to strengthen their own appeal to these volatile groups. They tapped into a vein of widely held anger at the humiliations and restrictions imposed by the Treaty of Versailles. Had the working class not been divided, the magnates would surely have been overwhelmed. Their salvation came from being able to play one side off against the other.

Driven by real fears that the working class left would seize control as they had in Russia, the magnates, like the aristocrats before the First World War, tinged their embrace of nationalism with imagined hostile intent in their neighbours. This permitted them to beef up the police and the armed forces and label all desire for change as sedition. Their media became more xenophobic, with a strong focus on German victimhood. They aligned themselves more and more with the fascists.

The ruling group which comprised the magnates, the aristocratic remnant and the embattled middle class seemed much stronger than the ragtag group of fascists that they used as a prop. They were confident they could control these thugs with the courts, the police, the government, the officer corps, the media and the factories. But this proved not to be so. The police and the courts, after a decade of making common cause with the fascists against the left were now so corrupted they favoured them as a matter of course. Fascist sympathy had spread into the ranks of the army leaving the officer corps unsure as to whether the rank and file would actually respond to orders to rein in the brown shirts. The middle class had become so bitter at impoverishment that they gave themselves over to the anger conjured up by the scapegoating rhetoric of a madman. The jingoistic press had become an amplifier for the fascist message.

If the magnates nursed a hope that the fascists would splinter amongst the various Freikorps leaders, they had not anticipated that Hitler would murder these potential competitors for power in the "night of the long knives." If they thought that a constitution would serve as constraint, they did not anticipate that he would torch the Reichstag as a pretext for dissolving government and bypassing these constitutional restrictions. He proved that he would do anything necessary to advance his goal of total control. By abetting the fascists during the previous decade, they had undercut the mechanisms they needed to oppose it.

Once the Nazis could no longer be controlled, the magnates had a choice. They could either put their factories, their power and their wealth at the disposal of the fascists or they could oppose them and have it all appropriated. They knuckled under, held their noses as they joined the party, and became an essential part of the fascist structure. The media was handed over to be turned into a propaganda machine for the party. Police officers and judges were required to be party members and the application of law become subservient to Nazi political goals. The officer corps was made to answer to Nazi political masters. The factories took instructions on what and how much to produce.

Nationalism, expressing on the one hand the exceptionalism of Aryan peoples, and on the other, the desire for revenge for the humiliations of the sabotaged war and the Treaty of Versailles, were the myths that justified all government activity. The Treaty of Versailles was publicly rejected; reparation payments were cut off; rearmament was taken up in earnest. This started the factories humming, funded initially by paying the magnates with secret bonds that were only to be redeemable at some indeterminate time in the future. The unemployed were soon back at work. Hyperinflation was ended with a controlled economy. These results legitimized Nazi rule and gained them more "soft" supporters. In Germany, the successor to aristocratic rule had appeared—a bureaucratic police state where the mass of workers were directed by a select minority who had been admitted to the party. There were echoes of Sparta here.

In the factories, the Nazis commanded both workers and owners, setting quotas, prices, wages and conditions of work. They demanded loyalty from the middle class and the professions, punishing those who resisted and rewarding with party membership those who went along. The party controlled the media, the courts, the army and the police directly as branches of the ruling party. They were brutal in demanding obedience from every sector of society. This melding of industrial capacity, the labour of the working class, and the skills of the middle class into a unitary structure created an effective state.

As before, to understand why this system worked and was stable we must look at the urges that drove the various individuals and the satisfactions that each could obtain. Nazi society existed as four distinct groups—the elite privileged leaders, the party members, the other citizens and the victims. The privileged leaders, many of them damaged survivors of the First World War, lived above the law holding the power of life and death over every other person. The Dominator and the Hoarder were unconstrained for them: that was the essence of their meaning and their route to satisfaction. They invented rituals to harvest expressions of esteem from the masses—songs and stories extolling their greatness, reflexive salutes and choreographed cheers. Everything made them the focus of these releases of shared anger and repressed fear. Experiencing the unrestrained screaming of tens or hundreds of thousands and being able to casually extinguish a life whenever it pleased them, made them feel like gods.

The party members, a privileged group that numbered up to about five million in a country of seventy million, occupied positions that allowed them to push others around, such as the police, the courts and the bureaucracies, so the Dominator was catered to, but the party also provided an encompassing group that gave meaning to their lives, gave them structure, and provided clubs and services for their welfare. The party also produced huge rallies designed to allow members to feel part of a larger entity. Their curiosity was circumscribed since esteem could only be garnered in ways acceptable to the party and to deviate from that was dangerous. However, with satisfaction available as the Dominator and with their encompassing group providing a sense of meaning, there were satisfactions for this group in embracing the regime.

Possibilities for satisfaction decreased for the other two classes. Citizens who were not in the party, the majority of the people in German society, had a more precarious existence. A party member might want their job, or their house, or their wife and, as a consequence, a non-party member could face trumped up charges in a rigged legal system. Legal protections were a charade; they had much to fear from attracting the attentions of the police, the SS or the Brownshirts for any reason. They lived between a fear of becoming one of the victims, and a hope of being raised into the party. Their labour, demanded from them for wages which were kept low, powered the rearmament machine so their existence depended on their usefulness. They had only the mutual aid of their neighbours, seriously undercut by the prevalence of informers, to lend them meaning and support. Their routes to satisfaction were very circumscribed. Enduring was their goal.

For the victims, the threats were real and ever-present, not just imagined and feared. These people provided the opportunities for party members to display their superiority through public humiliations and attack. Dominators need victims. Successes involved survival for one more day with the hope that the violence would fall on their neighbours and not themselves. This group, largely those designated as non-Aryan or those with a communist past, had few opportunities for satisfaction.

For the other governments in Europe as they faced a working class energized by the depression, the rising power and growing threat of the Nazis was a godsend. The unfolding drama in Germany gave them a focus for the nationalism and the rearmament they needed to prop up their own tenuous hold on power. Memories of the First World War made an active response to the traditional enemy an easy sell.

In all these countries, the Second World War, when it inevitably did erupt, gave the two strong classes that had emerged from the carnage of World War I—the industrial working class and the magnates—another chance to show who was dominant. It was obvious from the start that this was not going to be a reprise of World War I. Lessons from that conflict had been noted. The aristocratic leaders had been cast aside in favour of industrial planners. The advances in chemicals and machinery were now harnessed in new and more powerful ways. Germany led this advance towards a more technically sophisticated military. The others quickly followed.

It was crucial for any combatant to be superior in both equipment and logistics as well as manpower. Battles were won or lost with fleets of sophisticated airplanes, regiments of tanks coordinated as cavalry and naval armadas above and below the sea, all capable of carrying devastating firepower over great distances. These machines had to be supported with mountains of munitions and oceans of fuel, all available in the right amounts in the right place at the right time.

Those charged with organizing the production and shipping of these war materials, all drawn from the ranks of the magnates and the professional middle class, were given sweeping powers that began with planning production in the factory and extended through all the steps that connected the factory floor with delivery in the field. This coordinating power seated in bureaucracies mimicked similar arrangements that had already come about in fascist Germany. All countries moved rapidly in the direction of central planning as they worked to harness their industrial strength to prosecute the war. The Second World War saw a marked advance in state control over the economy in every country that was involved.

Companies that were able to quickly deliver large quantities of anything from boots to tanks soon found themselves operating with contracts that involved no risk, a guaranteed market which could never be saturated, one customer and assured profits. Competition was eliminated in the name of security of supply. Only companies that were already huge could fill these massive contracts. Big companies grew larger and wealthier, using their assured profits to swallow up smaller firms who were unable to compete because they were denied access to materials that were deemed essential to the war effort while also being too small to bid on the scale the war required. The Second World War restructured every economy into a world of fewer, but much bigger, corporations. This transformation occurred in America and Russia as well as Europe as they were all swept up in war preparations.

The industrial entities in America, Russia and all the countries of Europe grew so large and complex that one person could no longer comprehend the details nor direct and coordinate the activities of the tens of thousands of dispersed workers that were involved. The owners had to transform their firms into machines made up of detailed job descriptions that specified the tasks and responsibilities for each position as well as the varied communication channels that joined them to each other. Workers became simply predetermined cogs in one of these machines. If everyone performed just as was laid out in the plan, no more and no less, the procedures, rules and structures would make expected outputs appear. Companies ceased to be an expression of a magnate's will. They were turned into machines.

In Western Europe and America, neither the magnates nor the workers succeeded the aristocratic class to claim eminence at the apex of society. During the war, the two groups had temporarily set aside their differences to concentrate on victory. By the time the war ended, the workplace had evolved into a matrix of interconnected machines. Each group saw possibilities for gain in this development. The magnates, who worshipped growth, saw this as the route to bigger and better. The workers recognized that being needed as an essential cog in a machine would give them security and a lever with which to improve their position. Each feared jeopardizing these possibilities should they renew an adversarial relationship. Each needed the other for the machine to thrive. Each hoped that this arrangement would force the other to deliver their wishes. What emerged was a structure to which both contributed and from which both expected to draw benefits. The class war was over. The successor to the aristocratic order had appeared. It was the industrial machine to which both classes gave a fervent allegiance.

The workers offered their willingness to staff the factories and the military and, when massed and armed, to obey state instead of class loyalties. In return, the magnates would give the workers a larger share of the wealth generated by the industrial machine. They would pay a living wage and contribute taxes to allow governments to enhance the lives of all. Whether these taxes were removed as income taxes on workers, consumption taxes on consumers, capital gains taxes on owners or simply taxes on profits, it was all wealth that originated in the production process, wealth which the magnates had once kept for themselves. Both groups accepted the necessity for an expanded role for government, a hybrid creature tasked with standing between them to act as arbiter and organizer. This form of social organization became the norm, with slight variations, throughout Europe and North America. Russia and her satellites, though also evolving into massive bureaucracies, kept the government functions under the direct control of her new elites, thus melding the bureaucratic with vestiges of the aristocratic.

In the west, government grew stronger. It could soon deal as an equal with both the large industrial concerns and with the workers organizations. Both partners did all they could to influence government to favour their agenda but neither could actually direct its activities. The interconnected bureaucracies of the industrial sector, the unions and the government became the scaffolding on which modern society was organized. The twentieth-century thirty years war of 1914-1945 had finally produced a successor to the feudal aristocratic order: the world of the bureaucracy.

CHAPTER SEVEN
Life in the Machine

∿➤

Between 1900 and 1945, we transformed our world from a profusion of small independent wealth-producing industrial concerns where the workers, magnates and aristocratic elite jostled for primacy into a web of interconnected bureaucracies where activities arose automatically from a set of codified rules and procedures. This opened a whole new chapter in the human story. Feudalism was dead. In its stead, we had produced a world where everyone was simply a cog in a vast machine.

The leap to feudalism had released both the Dominator and the Originator from the smothering taboos and traditions of tribal life. As a consequence, for ten thousand years, power and wealth have been concentrating and technology has been growing steadily more sophisticated. These twin thrusts created both the industrial working class and the business magnates. The growth of these two powerful new groups gradually destabilized the existing power structure and led us to war. By 1945, when the aristocracy had been proven irrelevant, the two new claimants to power had combined their strengths to coax the benefits they expected from a bureaucratic industrial structure. This method of organization has been so successful in producing stuff that it has supplanted traditional feudalism everywhere.

Concepts, such as class, that had been central to understanding feudal societies became less significant in this new world while a group of other questions assumed critical importance. First, whose wishes were capable of directing this machine? If the lord in the manor and the boss in the corner office issuing commands are now largely irrelevant, then who determines the direction of change in this new world? Second, who owns this structure? At the very least, who draws benefit from its activity? Who does it work for? Third, why do we adhere to the machine with such fervent allegiance? What do we get out of it? Fourth, are there any threats lurking in this form of organization? Is this a good long term solution for humankind? And fifth, if threats do exist, can these be handled in ways that will lead us to a safe and stable future? In this chapter, we will examine these questions.

Perhaps the most important question for us is: whose wishes set the course for this world as it sails into the future? In tribal times, a coalition of men who had shown eminence all contributed to group decisions. In the feudal world, the strongest asserted themselves and the world then reflected their desires and their activity. Who is creating the future of the bureaucratic world?

A few centuries ago, businesses were small. They were directed by an owner with a few trusted aides carrying out his direct orders. The owner made all the decisions and then was rewarded or suffered losses as these decisions played out. His business was an extension of his personal will. When successful, he had profits to reinvest. Since it was natural to put his money into activities that he understood, businesses were impelled to grow. As long as markets existed, the number of workers could increase, output could increase, profits could increase, and expansion continued.

As these organizations grew, each worker had less and less contact with an owner who needed more and more aides to deliver his instructions. As his attention was divided, workers and aides would often be inactive as they waited for instructions. At some point, coordinating this group used up all the profits and expansion would cease. The size of any enterprise was self-limiting.

But profits needing to be reinvested were still being produced. So, in the 1800s, owners began developing more systematic methods of controlling large organizations. If a plan could be created with the contribution of each employee described in detail, then the owner's expectations would exist in the structure and he would not have to be constantly giving orders. Procedures were created which attempted to specify the appropriate action for every situation that an employee might be expected to encounter. Workers were recruited and trained to carry out these specific tasks. To make the training effective, the procedures had to be kept simple. Each job had to be made as narrow as possible. While the flexibility to carry out many different activities became impossible because no procedures were produced to cover those eventualities, the one expected thing could be done well. Jobs were joined together with pathways for both materials and information. These allowed each worker to know exactly what he needed to know, no more and no less, and have the materials that he needed to do his appointed task, neither too many nor too few, neither too soon nor too late. Management also became more mechanical by specifying which items were to be counted, which reports were to be filled in, who those reports were to be sent to, and what actions would automatically flow from the tabulated results. These techniques were effective. Firms grew larger.

But there was still a limit. At some point, the cost of counting inputs, filling in reports, and shunting resources would eat up profits, but that limit had been greatly extended by relying on a plan. During World War II, with guaranteed profits and with national survival demanding rapid expansion of the production of some products, producers were forced to push against these limits with urgency. Advances in electronics and communications, many driven by the search for more sophisticated weaponry, produced a group of communication and computational tools that made the organizational structures more efficient. Each new tool enabled more growth. The firms that emerged at the end of the war were many times larger and much more complex than anything that had existed at the turn of the century. With every increase in their size, they became more and more like machines.

These structures can only do things that have already been anticipated. Only then will there be procedures to initiate activity. These machines can only sense the world around them in ways for which there is already a report with a box to tick. Everything else ceases to exist for them. Access to resources that would be needed to respond to the unexpected had been pruned from the plan in the name of efficiency. Activity was limited to a small menu of possibilities. However, sensing one of these expected inputs would automatically initiate the corresponding activity. Everything was automatic. No one in the present seemed to be in charge.

There is a myth that the CEOs and other managers are driving these machines, that they are capable of devising and implementing responses to meet any unexpected circumstance. They claim to wield their companies as expressions of their will, just as the magnates did before them. But in reality they operate within a straitjacket of constraints—within the plans and procedures that define their firm, within shareholder demands to maximize short term returns, and within a matrix of contracts, laws and regulations that maintain the status quo. The only response to novel circumstances open to them is to create a new firm or subsidiary with a new set of procedures, structured at the outset to act differently. Incredible inertia sustains the status quo. Executives are driving a locomotive. It can pull a large load effectively down rails that were laid in the past but, if something off the track catches their attention, the whole structure must be rebuilt before it can respond.

Every firm is anchored in a web of competing firms, cooperating businesses, unions, suppliers, clients, government agencies and regulators, with the strands of this web defined by contracts, laws, customs and incentives, all of which have evolved over time to facilitate the core function of a firm, which is to increase its power and wealth. This hope for growth is behind every rule and procedure, and is the crucial feature of every strand in the web.

New structures with new rules are the only possible innovations. If this experiment generates profits, that node will grow; if it leads to losses, then the firm shrinks, their assets are transferred to others, and everyone learns from the experience. The structure as a whole benefits from these natural experiments to find the most efficient route to grow, allowing fresh ideas to creep in as it prunes the dead wood. Not only is growth at the heart of this world, but we are continually getting better at making it happen.

Any attempts to accommodate goals other than growth, such as environmental health, are stillborn as they fail the ultimate test of profitability. Executives, though limited in the goals they can pursue, do fill a role by keeping growth happening efficiently. Any other result ends with their demise. They are also cogs in the machine. They are not drivers. No one within the firm is actually in charge.

We often think that, as citizens and voters acting through our governments and civic agencies, we can advance a collective vision not based on growth. We hear this promise at every election. And if we compare the situation of anyone in the developed world today with that of an industrial worker in the 1800s, we can certainly see a history of significant gains realized by using government powers to oppose the magnates and their successors.

Government expresses our collective will in two ways—by restricting harmful activities with a promise of punishment, and by levying taxes and controlling the redistribution of that wealth. However, both activities can only occur after the fact. Things must have occurred before they can be taxed or before any punishments can be applied. Governments deal with consequences of activity, while the initial impulse to action remains untouched. The forces that are driving change remain inviolable within the procedures of the machine.

Citizens can certainly demand new restrictions, a new regulator or a new tax and this can have some effect on our lives. However, if we tried to deny growth itself to meet some other goal, the motive force of the system would disappear and we would soon be in the throes of economic collapse. There are areas that regulators approach very carefully, if at all. Government cannot steer the machine. It can only try to mitigate consequences after the fact.

Governments have tried to directly approach goals by establishing monopolies designed to deliver a specific service with no profit motive, funded through taxes and with everyone forced to become a customer. However, these agencies have always contained the seeds of their own eventual failure. Shielded from the possibility of pruning by protection from competition, delivering a service—military, education, health or the like—that causes distress when removed, and guaranteed resources by an ability to demand taxes, the people chosen to staff these agencies are given a secure base of personal power. This echoes the situation where a warrior was chosen to step above the group and the same dynamic ensues. As public monopolies become extensive, a new elite must emerge. Government agencies become fiefdoms. Because every person harbours the Dominator, governments cannot directly organize society by fiat.

Owners are not in control because of size. Executives are not in control because of the strictures of the rules and procedures. Governments are not in control because they can only operate after the fact. However, the lack of a controlling hand does not translate into a lack of change. Evolution does occur. Innovation leads to new entities with new procedures. Old entities grow, die or revise their procedures. The world of today is very different from the world of 1945. We do not need either a directing hand or a vision of the future to continually advance into new territory.

The quest for profits that is encoded into every procedure drives the machine forward. When particular businesses succeed, capital accumulates. This pushes the machine to grow. There are always a few who forsake short term profits to invest in ventures they hope will spawn new industries and pay huge dividends far into the future. And when stagnation looms, this trickle of money nurturing these innovations becomes a flood. New industries will appear quickly when markets start to saturate. No one plans any of this. Change always arrives as a surprise.

We animate the machine as we carry out our assigned tasks within our slot, but we are irrelevant to these societal changes. The more complex the machine, the narrower the job and the smaller our contribution to the whole. These machines move efficiently towards their programmed goals but there is no responsible agent behind this activity. No one is in charge.

Who Benefits: Let's Follow the Money

This machine is built to create wealth. It produces profits, goods and services in abundance. Someone consumes these goods, enjoys these services and benefits from the wealth. The machine obviously works for them. Who are these people?

At the beginning of the twentieth century, the robber barons in America and the industrial magnates in Europe exploited their workers and funnelled the profits into their own pockets. Both the beneficiaries and the victims were obvious. The workers were poor and had to fight for any improvements; the magnates were rich and in a commanding controlling position. But the growth of companies between that point and the end of the Second World War transformed both the position of the workers and the nature of the ownership class.

The magnates built the firms that brought railroads, electricity, industrial chemicals and oil to the world. Markets for these new products were huge. Labour was streaming into the industrial cities from the countryside. Growth was limited only by the amount of capital they could quickly bring to bear. So they tapped their moneyed friends, selling shares in their new ventures. Initially, the magnates maintained a dominant position in this group, but this soon changed. As the first generation of major investors passed on, their shares were divided amongst their children and grandchildren, associates and charities, spreading ownership more widely. Inherited shares could be sold to anyone. As continued expansion needed ever more capital, more shares were issued, now marketed to any stranger who had the money to purchase them. By the end of the Second World War, the ownership of most firms rested in a large group of shareholders, some of whom may have had significant holdings, though none was likely to be dominant. Like the aristocracy before them, the industrial magnates as a class had ceased to exist. Even if their heirs were wealthy, they no longer exercised any directing power over their organizations nor could they even claim the lion's share of the profits. The magnate's function had been replaced by the rules which regulated the stock market. The benefits they had been able to claim were now enjoyed by a large group of shareholders.

The shareholders, for their part, wanted only a safe and profitable return on their investment. To reduce risk, investors would purchase shares in many companies. If one performed badly, their income would be maintained by the others. They had no interest in directing any of these companies, let alone even scrutinizing the activities within them. Most companies were owned by millions of people with a legal claim to ownership and no interest in any of the details of operation.

These owners delegated the responsibilities of ownership to an elected group acting on their behalf. This group of directors were charged with hiring and overseeing the executives who ran the company. However, as the number of shareholders reached into the hundreds of thousands or millions, connections between shareholders and directors became more formal. Directors became another part of the company structure. Shareholder power was a myth. Ownership was only the right to a dividend.

A further change in the shareholder group occurred with the rise of mutual funds and pension plans. A mutual fund permits anyone able to accumulate a small amount of savings to purchase a share in the fund. The operators of these funds then buy and sell shares in companies to generate returns. The managers of funds have no interest in running companies—if profits drop, they sell their holding and move on. According to Wikipedia, at the end of 2020, total mutual fund assets worldwide were 63.1 trillion dollars.[1] Most of this money comes from individuals of relatively modest means as the truly rich will employ their own investment professionals and have little need for publicly available funds. The profits generated by the machine were making their way into ever more pockets spread ever more broadly throughout society.

Pension funds act in a similar fashion but the contributors are even less aware of the companies that support their income. Twentieth-century affluence in the developed world has permitted organized blue collar and professional workers to amass large pension funds which must be invested to generate the profits to secure future pension payouts. According to the Willis Towers Watson pension fund survey, pension funds in the twenty-two major markets had 56.6 trillion dollars under investment at the end of 2021 creating pension income for hundreds of millions of retirees.[2] These are huge sums.

Along with mutual funds and pension funds, a retail investment business has grown up to allow many workers and middle class professionals to own stocks and bonds directly or to maintain deposits with banks and similar institutions which use those deposits to buy stocks. Consequently, investments made by, or on behalf of, blue collar and white collar workers total well over one hundred trillion dollars. These are all claims on the profits of the machine. The very rich do still exist as privileged investors, but they are no longer dominant in hoovering up the cash. Ownership and its benefits have become widely dispersed.

The narrative of a small cabal of very rich people pulling strings from behind a curtain of wealth, though perhaps a somewhat valid description of the world of 1900, persists, even after these strings have been turned into fixed rules and procedures that can only be activated by the actions of hundreds of millions of ordinary people. Inequality definitely exists. Aspects of our feudal past, codified in the rules and procedures, guarantee this. The manufacture of inequality will get more attention later in the chapter.

So, no one is in charge and everyone benefits. Everyone drawing benefits is obviously a good thing. However, having change continuously emerge as a surprise can be a little bit worrying.

The Deal: Why We Love Our Machine

Government has transformed itself in the twentieth century. In traditional feudal systems, the state was simply an organizational tool which was useful to the elites in focussing the army, navy, laws, court system and police to help them exploit resources at home and abroad. When aristocratic power collapsed, the structures of government remained intact. Both workers and magnates turned to this neutral force, hoping they could control it in ways that would advance their agenda. It soon became strong enough that it did not have to take orders from either. Whenever there was a contest for control, compromises had to be crafted to keep the partners from erupting into violence. Government was where these negotiations occurred.

The working class agreed that state loyalties would supersede class loyalties, a promise which would make mob violence a thing of the past. This negation of the source of their power was the cost of their access to the goods-producing machine. The owners agreed to hand over in taxes money they had previously kept for themselves. This was the cost to them of a non-rebellious workforce. Government thus had a monopoly on force and large inflows of wealth. It grew rapidly, developing procedures to satisfy both the workers and the owners to some extent. It forged links to the worlds of money, management and ownership on one hand and to the worlds of labour, soldiers, voters and the citizenry on the other. Taking up this position in the centre of the web, it was the force that assembled the players into one big machine.

With each new function that government assumed, new rules were grafted onto the old agencies of control. However, the essence of the old structure remained. The core of the procedures was, as it had been for centuries, to facilitate growth and to ease the exploitation of both resources and people. The main change was who could claim the spoils. The aristocrats were out. Both workers and owners felt they were owed.

The workers claimed their share of the wealth. They toiled in a system designed to exploit them even as they enjoyed an increasing share of the wealth it produced. The industrial workers had arrived at the curious position of being beneficiaries of their own exploitation.

The workers pushed for larger wages and more entitlements and protections. The shareholder organizations and their firms used their position as employers, as owners of the media and through influenced politicians to work for maximum reductions in taxes and entitlements. Battles

7. Life in the Machine

within government over these issues kept the conflicts largely non-violent.

A continually shifting compromise evolved, swinging one way or the other as the relative strength of the parties waxed and waned. Laws punished union violence while forcing owners to respect legal strikes. Laws defined minimum wages, working conditions and job safety while taxes financed a growing list of entitlements. At the same time, government was vigilant in protecting the rights of property and assuring that profits could be made. Having government stand between them made the coalition work.

The coalition could continue to be successful as long as sufficient wealth was produced to provide a decent wage for the workers, a good profit for the owners, and enough taxes to support state activities. Luckily, during the few decades after the Second World War, total wealth grew so fast that everyone could get richer. No appetite developed to fight over the spoils. Paying workers a living wage unexpectedly created a new market. When workers could buy the goods they produced, it stimulated production which led to profits, which financed the higher wages which allowed more purchases. This cycle was important in driving wealth creation. The application of modern management techniques and the productivity gains that came from harnessing electricity, chemicals and the microchip in the factories also made large contributions to profitability. Steady increases in wealth kept everyone content.

One steel worker today produces about as much steel in an hour as a hundred workers could in 1918. Of this increase in generated wealth, a third to a half went to shorten the work week and the working years, retirement with a good pension now being possible after forty years of forty-hour weeks punctuated with holidays instead of a hazardous work life of sixty-five-hour weeks starting at age fifteen and lasting until the body wore out. Each worker also gained the spending power of ten to twenty workers of a hundred years ago, most of whom could never have dreamed of owning their own homes, let alone the style of homes, or the cars, toys, education, the appliances, vacations, clothes and cottages within reach of a steelworker today. The equivalent of the output of another ten to twenty workers funded entitlements like health care, universal education, courts, welfare and regulatory agencies through taxes on various aspects of the production process. And there was still enough left to assure the profits that would attract new shareholders.

When either workers or owners wanted to improve their position, they turned to the compromises that rested in government. These new demands could lead to new laws, new entitlements or new agencies. Or, when the balance of power swung towards the owners, it could lead to an erosion of these things. The machine became an ever more dense array of laws, regulations, traditions, structures, channels of communication, alliances, compromises, rules and procedures—some formal, some customary and traditional, and some others just understood as the way things were. Both workers and owners developed a strong allegiance to this machine because it worked for them.

Our great grandfathers and great grandmothers in both the industrial working class and the professional middle class enthusiastically embraced this deal. They agreed to be diligent in turning their job descriptions into reality. In return, they expected a living wage, a range of basic entitlements, and protection against abuses that had previously blighted their lives. Having just emerged from a devastating series of recessions and wars where they had been starved and used as cannon fodder, it was a no-brainer. The deal promised a secure life for them and even better prospects for their children. The deal worked. Their grandchildren and great grandchildren are living much more affluent lives than they could ever have imagined.

The deal is an all-or-nothing proposition. If workers carried out only half of the procedures or carried them out only half the time, the machine that fed them all would collapse as cog grated on cog. They assiduously carried out all of the rules all of the time. They agreed to take any grievance to government instead of challenging procedures directly. We have inherited this deal. It requires our devotion to the industrial machine.

As with most deals, however, the devil is in the details. And when we look closely, we see a serious threat crouching at the heart of this one. The basic aim of the procedures is to amass wealth and power. That motivated the feudal barons who initiated the structure; that drove the industrial magnates who adapted it; and that was the mandate given to the management gurus who wrote the rules that became the machine. Growth is the only goal. Procedures were carefully constructed to deny, deflect and eliminate everything that might interfere with growth. This paid huge dividends in a world with abundant resources. This was the dynamic that transformed the lives of the steelworkers.

But when we are faced with the limits of a finite system, unstoppable uncontrollable growth can be dangerous. Procedures, lurking in the structure and activated automatically, will drive us into areas that are best left alone. Our world is based on growth but, as we abut limits, it becomes clear that uncontrolled growth is our major problem. The deal, while making us safe and comfortable, carries within it the seeds of our destruction. But changing it seems impossible because of the depth of our devotion.

The machine is very good at producing and distributing things, many of which are available to us now and many others we can aspire to if we can only increase our salary. Our urge to hoard is constantly being stimulated and satisfied. Displays of clothes, houses and cars show our wealth in comparison to others. For the Hoarder, the machine is a paradise.

Wealth can satisfy the Dominator when we use it to force others to obey. Some jobs mandate that workers wield powers over others—we have all encountered bureaucratic tyrants who take enjoyment from pushing people around. With every satisfaction provided to the Hoarder and the Dominator in us, the machine justifies our allegiance.

Some positions demand discovery, design and problem solving and so provide opportunities to be the Originator in the service of a node in the machine. There are enough creative positions that everyone can aspire to achieve one themselves or to see their children grow into one. These satisfactions also confirm our allegiance to the deal.

Our need to feel ensconced in an encompassing group can also be met within work organizations. As the extended family, the village and independent local associations have faded away because of the need to wring out inefficiencies, our department at work can often be the only space where we feel part of a group. Our position—our cog in the machine—provides meaning. To be a doctor or a teacher or a manager or a longshoreman or an electrician in our workplace becomes who we are.

As we define ourselves in terms of the machine, the deal becomes so basic that it is not even seen. It echoes the parable of the fish recounted by David Foster Wallace in his 2005 commencement speech at Kenyon College:

> There are these two young fish swimming along and they happen to meet an older fish swimming the other way, who nods at them and says "Morning, boys.

How's the water?" And the two young fish swim on for a bit, and then eventually one of them looks over at the other and goes "What the hell is water?"[3]

The deal satisfies our need to be connected, to be the Originator, the Hoarder and the Dominator. Each of these satisfactions strengthens our allegiance. A connection that began with relief at being safe and fed, has become much more. It is difficult to question a structure that feeds us in so many ways.

So, no one is in charge, we all benefit, and we are bound tightly—and willingly—to this organization even though it may be a threat to our future. We had better take a look at this threat.

The Threat

Growth is central to the machine. It is not an incidental feature that can be adjusted by tweaking the procedures or by empowering a new regulating agency. Making profits to be reinvested is the reason the machine exists. This goal has shaped every rule, every procedure, every aspect of the structure. Growth is inevitable. Increasing the size and power of every node is the only measure of success.

This growth continually spawns more firms, bigger firms and more innovation. As this unplanned development produces problems, we respond through our elected bodies with new and larger regulatory agencies and increasingly complex structures of taxation and redistribution in the hope that more convoluted controls will smooth out the problems and allow us to grow forever. This busy work keeps us from recognizing that the problem is growth itself. We work hard at averting our attention because, not to do so, would force us to question the deal itself.

Building on a series of spectacular leaps in productivity and driven by huge reserves of capital that had been growing for centuries, growth since the Second World War has been phenomenal. The machine quickly grew much larger and became much more complex. More people could have their physical needs met and their need for meaning satisfied as they settled into their cog. The deal was constantly validated by this success. Fears for the future were belittled as being out of step with this reality. Workers became affluent, firms became too big to fail, governments morphed into behemoths, and innovations recast the world again and again. All political discourse—democratic, autocratic, communist, social democratic—assumed never-ending growth as the only goal worth pursuing.

Admittedly, there have been a few bumps in this highway to the future. Saturating markets caused growth to slow suddenly in 1929, just as it had in the 1870s. The world economy was heading towards collapse until rescued by the production of massive amounts of equipment to be wasted in the Second World War. This was financed by bonds which should have disastrously drained capital from the system when the loans came due, only magnifying the pain. But, two factors spurred growth so effectively after the war ended that, not only could the loans be paid off, but there was still enough capital generated to keep the pie growing.

First, radically refining the assembly line allowed firms to become much larger and more productive. These larger factories brought huge numbers of workers densely together and the massed numbers gave them a stronger voice. The unions they empowered forced both higher wages and an increase in entitlements. This wealth in the hands of working men gave them the spending power to pursue their pent up desires. As workers spent, they created demand for the products that their firms were producing. Production and profits were spurred to grow by the higher wages. Henry Ford, with the Model T in the twenties, had expressed the aim of producing a car that his workers could afford. This aim of a consumer society came to define the industrial system as a whole after the Second World War.

The second factor was a consequence of the first. Workers' dreams included a car so that they would no longer be forced to live near the factory gate. Instead, they could leave the tenements behind and move to a house with a yard on the edge of the city. This mass migration invigorated the automobile industry, the oil industry, the house building industry and the road building industry. These new houses needed furniture, appliances, carpeting, dishes, flatware, lamps, knick-knacks, everything. A new world was built on worker affluence. Each of these purchases generated profits, jobs and taxes. This transformation drove exuberant growth through the 1950s, '60s and '70s. However, all things run their course. By the end of the '70s, the pace was slackening because even these markets were becoming saturated. To sell a car in the '80s, you likely had to convince someone to upgrade. The same was true for houses, appliances and furniture.

There was, however, still growth that was based on a steadily rising population. In the 1950s and 1960s, the average number of children per woman in America was over three, reaching a high of 3.5 in 1958. These children—the so-called baby boomers—were entering their adult lives in the eighties, buying their own cars and houses. But this source of growth too was not destined to last. General affluence gave women a desire for equality, with the consequent opening to them of many previously closed educational and work opportunities. As horizons opened, the fertility rate dropped precipitously. It dipped below 2 in 1973 stabilizing a few years later around 1.8, a point well below the 2.1 needed to generate replacement in a population. From that point on, without significant immigration, the growth in population was destined to slow, eventually stop and then start to decrease. These changes heralded a reduction rather than an augmentation in demand.

Rapid growth stokes inflation as the more quickly growing sectors outstrip available resources and labour for a while. Inflation always favours monopolies over less strongly organized groups, investors and organized workers over those on fixed incomes or with no group to speak for them. It became a general feature of these decades of rapid growth.

All growth added to the demand for oil, the fuel which powered the industrial economy. Unfortunately, oil had only been discovered in a few inaccessible locations and all were ruled by autocrats. These producers were able to form a cartel and, in 1973, they forced a quadrupling of the price. Costs immediately jumped for everything that used oil in its production. Groups that could do so responded with immediate demands for price or wage increases of their own. Inflation really took off. Poverty amongst the powerless grew. Those who lived beyond the market, able to enforce their demands, got richer. This creeping inequality has added a darker hue to our picture of our modern world.

Growth constantly requires new frontiers. By the eighties, industrial activity was slowing again because of the changing demographics of smaller families and because the new world of suburbs and highways was nearing completion. The impoverishment of many by the oil shock was also leaching demand from the system. Some people bought less, some firms cut back, unemployment was on the rise. As we slipped towards recession again, money poured into the search for innovation. The potential inherent in the transistor was one such focus. This sparked the digital revolution. The building of the digital world then kept collapse at bay for another few decades.

Digital communications gave large swaths of the developing world access to both markets and contracts in the developed world. This has spurred an increase in wages for many and the growth of a middle class in countries like China, India, Korea, Mexico, Brazil and Vietnam. This new and growing middle class soon represented a huge market that had not existed before. This new market was significant in continuing to power growth.

When large swaths of people in these societies could see a possible route into the middle class for themselves or their children, they made the education of their children a priority. Except in those few countries where feudal attitudes still held sway, this included their girls. All of these women could aspire to goals other than motherhood. The fertility rate across virtually all of the developing world plunged. According to United Nations figures available from Wikipedia, the Total Fertility Rate—the expected number of children that a woman would have in her lifetime, averaged over the whole world—hovered around 5 from 1950 to 1970 and then began to drop, reaching 3.87 for the period 1975 to 1980, 3.44 for the period 1985 to 1990, 2.75 for the period 1995 to 2000, to 2.57 for the period 2005 to 2010. It is continuing to fall. This is a huge number of unborn people. As the article explained,

> ... a population that maintained a TFR (total fertility rate) of 3.8 over an extended period without a correspondingly high death or immigration rate would increase rapidly with a doubling period of about 32

years whereas a population that maintained a TFR of 2.0 over a long time would decrease unless it had large enough immigration.[4]

The same article gave the Total Fertility Rate for selected countries or regions in 2018 or, in a few cases, 2019. They gave an indication of where things stand now: India is 2.1, China 1,6, Japan 1.42, South Korea 0.92, Brazil 1.75, Mexico 2.22, Europe 1.55, North America 1.71. Some countries, like Japan and Russia, are already declining in population. China is peaking now and will decline in population going forward.

Population increases in most of the developing world are now being driven by a lengthening of life spans rather than by an excess of births over replacement. This impetus is decreasing as significant gains from extending basic health care into most regions of the world have already occurred. It is true that high fertility rates are still the norm in some regions of Africa and the Middle East that are so mired in poverty and autocracy that the creation of a middle class and the education of girls has been prevented, but there are signs of change here and these regions can realistically be expected to mirror the rest of the world. Given these evolving conditions, The United Nations Population Dynamics group have generated population predictions which show a peak population of around eleven billion people in 2100. Their models, however, are acutely sensitive to changes in fertility rate and have been criticized as being unduly pessimistic. If the birth rate should drop slightly more quickly than they anticipate, especially in a few crucial countries such as Nigeria and Ethiopia, then a much lower peak would occur much sooner. Jorgen Randers is a Norwegian academic who co-authored *The Limits to Growth* in 1972 with its warnings of exponential population growth. Since then the numbers have induced him to change his mind. His current thinking is that "the world population ... will peak at eight billion people in 2040 and then decline."[5] Other demographers are reaching similar conclusions. The Club of Rome, the group that initially commissioned *The Limits to Growth*, released in 2023 the results of a follow-up study to their original apocalyptic warnings. It projects that we will have a high of 8.8 billion people on Earth before 2050, followed by a rapid decline.[6]

A 2014 article in *The Economist* titled "Don't Panic," noted that the United Nations failed to forecast "the spectacular declines in fertility in Bangladesh or Iran since 1980 (in both countries, from roughly six children per woman to about two now)."[7] In whatever fashion the details play out, it is clear that a worldwide slowing of population growth has been triggered by a spreading affluence that is due largely to global connectivity. This must drastically affect the size of future markets for the industrial machine.

Every state will soon need significant immigration just to maintain the size of their internal markets. Obviously, all cannot do this in a world where the total population is static or decreasing. It is most likely that growth will continue in the developing world as recent immigrants choose to return to their country of origin, using their attained experience and wealth to establish themselves in privileged positions. It has been noted that, in periods where economic prospects in Mexico improve or times in America get tougher, net migration across America's southern border tilts towards Mexico.[8] There is a strong pull towards home. The diasporas of India, China and Latin America may flow more strongly towards their countries of birth, invigorating economies there while leaching even more demand from economies in the developed world.

Many developing countries also still have some rural agricultural workers who would like to join the middle class. Increasing the mechanization of agriculture would liberate these workers. This reserve of people joining the market economy can keep their markets expanding even as total population may be decreasing. Such mechanization will also increase the production of food. This drive to urbanization will continue the downward pressure on the fertility rate.

This demographic change from growth to decrease will be an existential challenge for an economic system based on growth. A state with a declining population has no need of a house building industry—there will be enough houses for everyone, and some will still be sitting empty. This is the situation in many Japanese cities already.[9] We will need fewer cars, appliances and furniture, causing used markets in these things to saturate, further undercutting the demand for new production. Export markets cannot be a solution as every country will be sharing the same challenge. The need to service an increasing population, which has been a powerful driver of our economies for two centuries, is about to disappear. We need an economy that can be adapted to a steady state rather than continual expansion. One with growth at its core will be a problem.

Even if there can be winners in the initial competition for immigrants, this will never be more than a delaying stratagem. Total global population is going to grow more slowly, then stabilize, then start to decrease. Significant population flows will occur in response to differences in opportunities between states until those differences start to even out. Then immigration flows will become a trickle. Every state will have to face the challenge of basing an economy on a declining population. When parts of their market disappear, firms must shrink. Jobs must disappear. At that point, unemployment and poverty will return. Tax revenue will plummet. We are never far from the spiral of collapse.

Hard times always threaten equality. Though humans are loyal to those in their tight little groups, this fellow feeling does not apply to strangers. We have a natural inclination to see the world in terms of us versus them. Their resources tempt us. Their strength is seen as a threat. We scapegoat them to explain our failures and unhappiness. Pogroms and wars are the result. The Dominator thrives in hard times.

The economic incentives offered to scientists, engineers and innovators to uncover new industries will be huge. Change will come at us faster than ever before and from many quarters. Such desperate change always carries a real possibility of unintended consequences. The larger the amount of capital propelling the change, the harder it will be to assess the inherent dangers before they are upon us or to deal with unintended consequences when they do appear.

This is not a soothing scenario. Either we can expect destruction from unintended consequences arising from a desperate attempt to stimulate growth in the face of a declining population or no appropriate innovation will come to our rescue, the economy will continue to shrink, profits will disappear, firms will crash, unemployment will soar, desperation will spread and violence will follow. The threat lurking at the heart of the deal is real. It is approaching fast. Absolute declines in population will be here in thirty to sixty years. Now is the time for us to make changes.

The Hope

As dire as this all may be sounding, we are not helpless. Human beings are complex and flexible creatures. We have surmounted crises in our past by reorienting our survival strategy so that it gains strength from a different mix of motivating urges. We can do so again. Our situation today is unique in a few significant aspects. First, the community facing the challenge is much larger than a few tribes, a village or a small city-state. Global connectivity means that between eight and ten billion people are facing this threat together. Secondly, we have developed a body of knowledge that allows us to contemplate the future and perhaps forestall collapse before we are engulfed. Alternatives can be tested in thought experiments and may not have to be attempted during collapse. We are engaged in a race between the consequences of relying on the Dominator which promise to lead us inexorably to destruction and the outcomes of relying on the Originator which give us the ability to anticipate and overcome that same predicament.

Remember that the machine has no one in charge. There is no overseer with sword and whip choosing a goal, forcing us to act, actively creating these problems as we fulfil his dreams. Instead, the script emerges as a surprise from instructions which have been encoded in the past into rules, procedures and structures. We comply with its demands because we are strongly invested in the deal. But other futures are possible. We can make them happen.

As an example, the Hoarder in us asks that we always seek the lowest price. We drive firms to exploit in order to lower their costs so that they can satisfy our demand. Then, as citizens, we establish regulators to police this exploitation. These activities define much of our world, but it all starts with our individual decisions in the marketplace. No one forces us to act like this.

Whenever the machine senses an input, the corresponding activity is triggered. Period. No argument, justification or thought is either necessary or possible. Stimulate the input and the prescribed activity will occur. By identifying the inputs and consciously controlling what they sense, we can make the machine work differently.

We generate most of those inputs. Our demands in the marketplace initiate chains of instructions that connect the extraction of raw materials to the delivery of a product into our hands. Our demands in the political sphere direct regulating and redistributive agencies of great power. We, as workers, push the buttons that carry out the procedures. And some of us are even tasked with fashioning the new rules that will create new parts of the machine. We have tools to steer this machine.

The rules have been designed to attract the Dominator and the Hoarder within each of us. But human beings are more than this. We are also Cooperators, an aspect of human beings that has been suppressed during feudal times. And we are Originators, driven to follow our curiosity in ways not necessarily mandated by a search for profits. These two drives, if respected in our societies, can lead humanity in very different directions. And that is where we are heading in Part II.

Part II

The Future

Chapter Eight
Taking Control

The deal provides benefits. In just a few centuries of industrial growth, we have advanced from a world where 90 percent of people lived with food insecurity to one where less than 10 percent are hungry, despite a massive increase in population. The industrial machine has improved our clothes, our houses, our comfort, our mobility, our ability to communicate, and our health. As markets for stuff saturated, the machine shifted to producing experiences and a vibrant service sector now adds much to our lives. Any challenge to the deal must protect these gains.

The machine will try to grow out of every problem by incentivizing new forms of innovation. The previous few decades have been driven by a desire to make digital storage and communication widely available and extremely cheap. The result is that people all over the world have become connected to each other, all people have been connected to knowledge, and every continent is connected to the others. These changes have been transformative in the sharing of wealth and in defusing the population bomb. This connectedness provides, for the first time ever, the possibility of global cooperation.

The climate crisis is now forcing investment to flow into a variety of new energy systems. It is likely that the cost of clean energy will follow the cost of communication and tend towards zero as breakthroughs and price reductions occur in fusion, solar, wind power, batteries and other energy technologies.[1,2] This promises to open our way to new tools that can undergird entirely different ways of being in the world.

Investments building on the reading of the genetic code are revealing previously hidden biological truths. These are translating into other tools that will reshape our future. Unfortunately, all this research is being directed by the machine and driven by the quest for profits. Those applications that generate profits see the light of day—all others become unmet funding requests. We must assure that these innovations further our happiness rather than a bank account in some node of the machine.

Change will occur. Can we direct this activity towards outcomes like freedom, responsibility, clean air, habitat conservation, climate sustainability and peace? Can we maintain the productive capacity and innovative energy of the machine while directing its energy towards meeting these other goals? That is our challenge today.

Responsible Consuming

We produce enough stuff to easily meet the needs of everyone on the planet. We have no need to produce more. This is true now and will become even more true in the future as communication and energy innovations make production more efficient and as population stops growing. And yet, even as production continues to increase, need amongst the poor remains a persistent feature of our world. And even as producers become more efficient, our lives are more and more dominated by work. Things obviously can be different.

When a firm senses that we want something and that we have money we're willing to spend on it, that firm begins to organize production and distribution. Everything follows automatically from our articulation of the demand. But the poor have no money. They create no demand. Producers cannot see them. From the viewpoint of the machine, they do not exist. But for those with money, every whim generates activity.

The buyer is in control. Producers, needing our business, must be attuned to our wishes. This can generate real consequences. If most people forego a product, it disappears; if many express a need for a new product and a willingness to spend to get it, it will soon become part of our lives. The marketplace is a reflection of our wants, our demands and our activity.

If everyone chooses their purchases by the light of their beliefs and hopes for the future, then the world will soon reflect those beliefs and hopes. On the other hand, if we act as if we believe that every item we buy has no history, no importance and no wider consequence, that all that matters is attaining the lowest price, the most attractive packaging or the slickest advertising, then we force these features to become defining characteristics of our world.

Every purchase we make carries a message. When added with similar messages from others, it will trigger a chain of activity. When we choose brand A over brand B, we are calling for a world in which the maker of brand A will thrive while the maker of brand B should change, shrink or disappear. We have helped to create our future with that choice. The reasons for our choices, when we make them known, become features on which producers can compete. As an example, if enough people make it known that they will buy only from producers who pay a living wage, even if that results in a higher price than that of some competitor who relies on sweat shops, then that message will condition the behaviour of a group of producers. They will increase the pay cheques of their workers and then try to make us aware that they have done so. As more people share in making this demand, more producers will have to pay a living wage. If most people come to believe that exploiting workers is wrong, and back that belief with their purchasing, then sweat shops, child labour and forced labour will disappear. There is nothing stopping us from doing this but our fierce adherence to the deal.

When we base our purchasing decisions simply on the lowest possible price, we are demanding a world that is defined by a race to the bottom. We are forcing producers to do everything in their power to deliver the least expensive product because that is the only way they can get our business. They must pay workers as little as possible, demanding maximum effort and compromising worker health if it promises to lower production costs. They must offload costs into the commons by evading pollution abatement rules because it will raise their costs not to do so. They must hide these facts from us so that we won't see ourselves as complicit in the problems we deplore. However, no matter how successfully the intervening stages may obscure the link between our purchase and the social ill, we alone are the responsible parties because it is our demands that cause it to happen and we alone have the power to change it.

Making responsible choices is difficult when we do not have trustworthy information about the production and delivery of our products. That information is now hidden,

8. Taking Control

buried in the arcana of the machine and distorted by image advertising. But a way to provide this data is now feasible. A trustworthy independent firm could gather this information, verify its accuracy and make it available on an app that associates every product with a report card. Scanning a code on the product would bring up a screen with a green check or a red cross for each issue on which we seek direction. This would give an ethical picture of the product immediately. Drilling down on any issue with a touch of the finger would give us more detail. The full reports would be linked should we wish to investigate further.

Firms are incentivized to whitewash their image. Governments are reluctant to offend their allies in ways that may impact tax revenues. Therefore, this agency must be completely independent of both industry and government. If we want it, we must pay for it. However, hundreds of millions of subscribers could easily support a robust investigative presence. A group of such agencies competing on the issues of credibility and presentation would be ideal. We will return later to the creation and control of necessary tools.

If we make it known that we are prepared to pay for it, then such an agency will soon exist. That is the beauty of a competitive marketplace—it translates stated needs into reality. When we make it known that we will only make purchases that reflect our values, responsible firms will soon appear. Any firm that wants our business will have to welcome the scrutiny of this independent agency. They would have to look at their practices through its lens and the eyes of the subscribers it represents. Producers would have to start competing on criteria in which we had declared an interest. These firms would have to truly reflect our beliefs. The business decisions and the redesign of procedures would be theirs, but the ultimate directing impetus would be coming from us. Allocating our resources with intent has power.

In the marketplace, we don't all have to agree before activity can occur. One person expressing a unique desire is an eccentric who will be ignored by everyone. A group of people voicing a shared demand, however, move beyond eccentricity into becoming a niche market which will be able to support a small firm or induce an established firm to add a new product line. The internet enables small and widely dispersed groups of people to form such niche markets connected by online sales and home delivery. As the number of people voicing a particular demand becomes larger, more and more producers will change their behaviour in response. No committees must be formed to decide on a best option. No compromises must be found to forge a way forward that is acceptable to all opinions. No products must be banned. Change will grow and spread as a reflection of the change occurring within people.

If half of the people want A and the other half want B, then the market easily supports parallel sets of producers. Both groups can have their needs met without having to deny the needs of the other. In the marketplace, difference leads to variety. The market creates myriad solutions as different producers respond to our demands in their own fashion.

Our role as active consumers is to understand what is important to us, to articulate clear demands, and to be willing to support the firms that meet these demands. In that world, consumers drive the machine. Currently, we are driving blind, investing our energies into complaining when our blindness causes accidents. It is a colossal waste of time and energy for us to blame producers for responding to our demand for cheapness and then begging governments to save us from ourselves.

Our purchases should advance the demand that no one be exploited in the production process, from the underclass in the developed world to the sweat shop worker or indentured labourer in the developing world. Everyone in the world should be able to expect a living wage for labour, a wage which will allow them to eat well, to be warm and housed and comfortable, to be secure and able to raise secure children. We can attack the problem of poverty in the midst of plenty that is created by our insistent demand for cheapness. Just as happened after World War II in the developed world and in the last few decades with the growth of the middle class in the developing world, economically empowering marginal workers will boost aggregate demand, and this will benefit everyone. Buying responsibly is not a zero-sum game where someone must lose before someone else can gain.

If our purchases carry the demand that damage to the commons is unacceptable, then the price of many goods will increase as firms become diligent about eliminating pollution as part of the production process. In this way, we will be purchasing clean air, clean water and respect for ecosystem integrity. Articulating beliefs and having them direct our purchases can affect every aspect of our world.

We don't just have to limit our demands to influencing those things we currently consume. The marketplace reacts to inputs. If we can shape a desire into the form of a result we are willing to pay for, and if we can make that willingness to spend clear, then firms will arise to meet that demand. For example, if we value oceans without plastic and decide that we are prepared to pay for that, then a variety of services will arise, competing on cost and effectiveness, all trying to find better ways to gather plastic, dispose of it or repurpose it into new products. A desire for less carbon dioxide in the atmosphere can connect individuals willing to pay for this with firms that will capture carbon and either store it or repurpose it. One person is helpless in the face of these problems, but billions of people with a tool that can aggregate their demands will be able to accomplish much quickly.

We currently pay taxes to fund efforts in these areas. However, the need for compromise and unanimity before any politically organized action can take place and the sensitivity to criticism from pressure groups that is a feature of the political process, cause most of the energy to seep away from the goal into activity in the political process itself before action can begin. Deploying our resources ourselves connects our feelings of responsibility directly to efforts aimed at achieving the goal.

Active Citizenship

We also have power as citizens that is currently extremely underutilized. We emerged as human beings when we became able to cooperate in large groups. This human superpower was opposed by feudal elites who correctly saw empowered groups and self-organization as threats to their power. When the Industrial Revolution unexpectedly empowered the masses, wider suffrage had to be conceded. This franchise was granted in ways designed to placate rather than empower the citizenry. The much sought after vote was reduced to a ritual which allowed people to signal periodic acceptance of their rulers. This passive citizenship denies people the ability to turn their many specific beliefs into action.

Active citizenship would consist of tools to focus the joint activity of many on specific projects. We should be good at this. Our civic structures should coordinate efforts to deal with common concerns that are not amenable to solutions through the marketplace. We have the power to change our civic activity from passive into active.

As government agencies became powerful parts of the machine, their instinct for self-preservation led them to erect barriers between themselves and the people they ostensibly served to protect their power from scrutiny, criticism and control. They developed jargon which only they could understand. They made their rules and procedures obscure to the uninitiated. They deflected inquiries into public relations departments expert in the arts of doublespeak, whitewash and hogwash. Opportunities to connect individual feelings of responsibility with activity were designed out of the system. The thrust was always to reduce civic engagement to the ritual of the infrequent ballot. Citizens were expected to pay taxes and obey instructions.

We need agencies that can channel our beliefs into activity. We must be able to connect specific beliefs, efforts and payments with expected outcomes. As we experience the results of our choices and our activity, we must be able to evaluate and make changes. We can have civic structures that are living, growing, changing, active agents in the world. Discussing, deciding, defining, paying, experiencing, evaluating are all active verbs—this area of our lives must be active throughout.

In those rare instances in history where civic life rested on active citizenship such as in ancient Athens, the medieval city-states and the independent towns of early America, cooperating groups of people could create a vital civic life. These groups made decisions and contributed resources to solve shared problems. Groups were small enough that people knew others by character and by reputation: when someone spoke, their words carried weight if they had earned respect, or they were discounted if they had shown themselves to be insubstantial. The need for compromise encouraged open-mindedness. Neighbourhoods were alive with discussion because every voice and every choice mattered. Everyone knew they were creating the future they all shared. This active citizenship can occur when people control their beliefs, their resources, their decisions and their actions. Cooperating is natural to the human animal. If we are empowered, we will join with others to meet shared needs.

As was true in the marketplace, we can create a responsible civic world when we are willing to stand behind every action we demand. Just as our purchasing decisions initiate a chain of activities in the marketplace, so does the expenditure of our taxes in the civic sphere. This civic activity is done in our name and financed by our money. We voice the demands so we, not the politicians or the bureaucrats, are the responsible parties. To be responsible, we need to know the effects that our money and our demands are causing. Government assessments of their own actions are self-serving and not believable so, as in the marketplace, we need accurate independent information about the effects of our resources. This could be provided by extending the scope of the agency established to evaluate the marketplace or it could be provided by a parallel agency. Either way, we need an accurate picture of both the problem we are trying to solve and the impact of our efforts. With this, we can choose to fund tasks to carry our beliefs into the world.

The route to a responsible world lies through myriad acts of conscious consuming and active citizenship. Consensus will emerge from the summation of many individual decisions to allocate personal resources within a marketplace of options in both the civic and commercial spheres. However, marketplaces have shown that they have two inherent weaknesses. First, some firms will respond better to new demands. These winners will grow stronger over time and their size and strength will come to offer them systemic advantages. Power tends to concentrate into fewer hands which then helps power to concentrate further. And secondly, the competitive aspect of a marketplace creates individual winners and losers. This exacerbates inequality between individuals. We will look briefly at these two effects.

The Corrosive Effects of Concentrating Power

An economy based on individual decisions to buy and sell—capitalism—concentrates wealth. In every transaction, the wealthier have an advantage. They can outbid, wait for advantageous moments and monopolize opportunities. Wealth and power will concentrate in successful nodes in the machine. Every sector of the economy and every branch in government will tend towards domination by the successful few large entities. Choice will be reduced as we must accept what these monopolies deign to offer. Our demands will no longer drive the system. The end of this road is one big conglomerate.

In a truly competitive world, opposition to wealth concentration would come from creative destruction: as new firms arise and outcompete the old, the old must break up and fade away. Competition here would result in the growth of the new instead of the domination by the few. In this world, power would be continually redistributed rather than concentrated. This mechanism does indeed exist, but it fails when firms become large enough to control resources, markets, technology and the entry of new firms. At some point, one or a few firms will cross this threshold and become "too big to fail". Creative destruction is only effective in immature markets. It is a postponement, not a solution to concentration.

The scaffolding for modern capitalism, such as banking regulations and contract law, is manmade. However, these rules have always avoided addressing this tendency to giantism because to do so would be a direct challenge to growth rather than an attempt to shape it. But a healthy marketplace must include a counterweight to wealth concentration. One potential tool is anti-trust action brought selectively against the monopolies that cross the threshold. This has had negligible effect. The selection process is arbitrary and dismemberment only turns a monopoly into an oligopoly which is just as effective at stymying choice. Meanwhile, power continues to concentrate.

However, there is a simple strategy which would be effective. First outlined by the American journalist and reformer Henry George in 1880 in his book *Progress and Poverty*, it has been well known and regularly ignored for a century and a half. He was specifically addressing the monopolization of land where, in his time, a few individuals controlled massive tracts and collected rents while most people were without homes and powerless to apply their energies to alter their situation. However, his idea is relevant anywhere that financial power tends to concentrate. Soon after its publication, over three million copies of *Progress and Poverty* were bought, exceeding all other books written in the English language except the Bible during the 1890s. By 1936, it had been translated into thirteen languages and at least six million copies had been sold.[3] The idea was obviously well-known and it resonated with many people.

In any sector—oil, cars, soybeans, copper, operating systems, whatever—concentration can be directly discouraged by creating a progressive tax disincentive to grow. Suppose that the tax rate that is applied to a firm rises with the percentage of the market they control. For example, if a firm has less than a specified share of the market—say 5 percent—their tax rate would be a modest A percent. For firms controlling up to 10 percent of their sector, their tax rate would rise gradually to become twice A. For firms controlling an even larger share, the tax rate would continue to rise towards three or even four times A making it increasingly difficult for larger firms to compete with the smaller. All that would be needed would be to choose the number reflecting the degree of market dominance that squelches competition and the tax rate that would disincentivize growth for each product category. Such a system would be easy to administer. Companies now file financial results and pay taxes based on various metrics so it would require no more than a change in the way that taxes are assessed based on their filings.

This denial of dominance would continually encourage new players with new ideas. Most firms would come to exist around the point where further benefits from an economy of scale are outweighed by the increasing tax burden. We can continually adjust this point to assure a competitive marketplace. Such a system would value variety and innovation rather than growth. Companies would have to compete by becoming more attractive, more varied, more efficient, more useful and more responsive to their customers.

Banishing Inequality

The second corrosive consequence of capitalism is its tendency to widen divisions between rich and poor. As this tendency proceeds, the basic needs of many must be met with redistribution. This fosters both disempowerment and resentment. Cooperation becomes impossible because each class inhabits their separate communities of donors or takers. The constant interaction of wealth and poverty generates fear, vigilance, envy, violence, disempowerment, poverty, isolation, ugliness and casual cruelty. Community governance becomes a matter of suppressing the inevitable violence that arises from the coexistence of two solitudes. Inequality is a disease on the body of society. Growing inequality is a cancer that must be cured.

On the other hand, equality is the magic elixir that generates cooperation. If there is no "other" in our world, then actions that enhance all lives will be given freely in the certainty that others will respond in their turn. As the examples of ancient Athens and the medieval city-states made clear, widespread equality induces activities which minister to our shared space. We all gain when our community becomes more beautiful, more safe, more prosperous, more comfortable and more vibrant. With equality, our community can become a valued part of ourselves. It is a tragedy that the marketplace, in granting freedom of choice, sows the seeds that can overwhelm this Eden.

This inequality displays as poverty on one hand and excessive wealth on the other. It may have been essential to the growth of the industrial world to have a class that would take on the dirty, unpleasant, dangerous, debilitating jobs that have built our world. Only their crushing fear of poverty and starvation made people move into the mills, the mines and the slums. As the wealthy got rich off this suffering, their accumulating capital and their drive to gain wealth initiated the changes that transformed that world into our own. But we no longer need suffering to power change. Instead, it is now an impediment to a healthy society.

If all workers are assured a living wage, then people will choose to work at dangerous, dirty jobs only if lured by positive inducements. Larger pay, rather than smaller, would attach to these positions. These labour costs would then stimulate ever more innovation to reduce the dangerous and dirty aspects of the jobs. We are a rich society. A few people are very wealthy. A small group remains in dire poverty. The majority live comfortable lives. We can afford to move against this scourge.

We can argue that eliminating poverty is the right thing to do. It is irresponsible to turn away from suffering when we have the resources to relieve it. And this is undeniably true. But self-interest should also be a powerful motivation. Eliminating poverty is a prerequisite for a vibrant community.

Imagine a world without poverty. Visible inequality inflames envy and anger, setting the stage for most crimes. Eliminating poverty would make crime a rarity. When crime is rare, police would become a small presence of helpers rather than a large cadre of enforcers. Prisons could become a tool focussing on overcoming the barriers excluding a few people from society rather than being warehouses for huge populations of offenders.

Many children in poverty suffer from inadequate prenatal and early childhood nourishment which blights the rest of their lives. Human development cannot be optimum when the proper building blocks have been denied. Inadequate nutrition for a particular week, or month, or year disrupts development, condemning people to be always less than they could be. Beyond nutrition itself, epigenetics plays a part. For example, if the developing body senses a dearth of food, then versions of genes become active that cause growth to be attenuated to fit people for a sparse future. This flexibility has helped the species survive, but creating such conditions purposefully weakens individuals and, as a consequence, their community. Imagine a world where everyone could grow to their full potential.

Poverty generates stress. Fear of having your home taken away and the lurking fear of food insecurity can never be ignored. This fear will display as frustration, self-blame and anger, erupting as conflict between husband and wife and between parents and children, yielding a harvest of family abuse and traumatized children. Turning to the Dominator in these extreme situations is a very human thing to do.

Such trauma in the first few years is yet another powerful epigenetic trigger which promotes enduring characteristics such as reduced impulse control and an inability to empathize with others, traits useful in a world of random violence but unsuited to successfully navigating the modern world. Evidence strongly correlates early childhood trauma with the creation of bullies and violent and addictive behaviours later in life.[4] Imagine if this violence was eliminated at the source. Chronic stress also undermines physical health, weakening the immune system, promoting heart disease, undermining working memory and decision-making skills, interfering with digestion, and inducing depression. A reduced life expectancy for the poor displays as a myriad of ills, all tracing their roots back to poverty. Eliminating poverty would banish worlds of pain and suffering, both for individuals and for society.

We can afford to guarantee everyone a basic income provided with an assurance that it can be counted on into the future. Such a guaranteed basic income would not be technically difficult to arrange. Imagine that a taxable payment sufficient for basic living would be given to everyone in society with no needs test and no strings attached. At the same time, the basic income tax rate for everyone would be raised to a level such that any individual making a set amount—say $60,000, $70,000 or $80,000 a year—would have the same after tax income as they would have had before.

For those people, the increase in their income due to the universal payment would exactly match the increase in taxes they were being asked to pay. Everyone who had previously made less than this amount would have more money in hand because their extra tax payable would be less than this new payment; everyone who had previously earned more would have less in hand because the increase in their taxes would exceed the extra basic income payment they received. Designing the system would only involve setting two numbers—the amount of the basic income needed to banish poverty and the level of income where the gain from the universal income and the increase in tax match each other. These two numbers will then generate the rate of tax that would apply to all. Approaching the problem of poverty directly like this will yield the success that can never arrive from indirect solutions such as waiting for prosperity to trickle down from the rich.

However, this is not a zero-sum calculation involving only a transfer of taxes from the richer to the poorer. The scheme itself will produce savings that will contribute to the cost of the universal payment. Such a guaranteed payment—given to all with no means test, no application procedures and consequently no need for enforcement against abuse—would replace many existing transfers: old age pensions, disability pensions, veterans allowances, welfare payments, sales tax rebates, children's allowances, student support programs, grants to artists and the like. The administrative and enforcement costs of all of these programs would also disappear. When everyone has a living income, we will no longer need Band-Aids. We would also save much of the money we spend enforcing poverty with policing costs, court costs, prison costs and health costs. These savings would continue to grow as fewer traumatized children were launched into the world and as the long-term health consequences of living with constant stress decreased. Other benefits would also naturally accrue from reducing the fear, violence and envy that is nurtured by inequality. For example, less money would have to be spent on insurance, security, fences and guns.

These reductions in direct costs would be significant gains but a larger benefit would come from the expansion of the middle class. This guaranteed payment would be in addition to any wage people now earn. For the working poor, this topping up would change their lives from eking out an inadequate living to actually having disposable income. This would translate into increased demand for many items, stimulating economic activity. The rich, as well as the poor, would share in the growth of this pie.

One argument for the preservation of poverty has always been a belief by some that, without a credible threat of starvation as a stimulus to work, people will just sit around eating chips, drinking beer and watching television. Evidence from the few universal income trials that have been attempted have indicated that this is patently not true.[5] The basic income would be basic—enough to allow survival. Earnings on top of this would be not only allowed but encouraged. Even though these earnings would be taxed, it would be at a sufficiently low level that these earnings would make a big difference in people's lives. These earnings could mean the difference between living in a room and living in a more comfortable apartment, between basic nutrition and going out to a restaurant occasionally. There is so much that can be added to one's life that a motivation to earn would be undiminished by a guaranteed basic income. There is little diminution noted in the desire of well-paid people today to earn more so they can spend more on luxuries and continue to accumulate even though they are in no way motivated by starvation.

It is frustrating to note the litany of ills that have resulted from the simple fact that we have not evolved mechanisms to oppose capitalism's tendency to concentrate wealth. We have been passively waiting for this to happen for far too long. This lack has left a collection of self-inflicted wounds on the body of society. There are huge benefits to be gained by rectifying this omission. The list of "intractable" problems that trace their roots back to poverty is long and depressing. They can be made to disappear. Active citizenship means identifying problems and implementing solutions that reflect our beliefs. This one is within our grasp.

Even though a universal basic income could reduce poverty, it could not, by itself, create equality. The gulf between the very wealthy and the rest of society would remain. Our society would still contain two separate worlds. The increase in income taxes needed to fund the universal basic income might be a small bother for the wealthiest but that loss would soon be offset by the stimulation to their firms and their investments from the expansion of the middle class. To reduce inequality itself, capitalism needs another mechanism to direct accumulations of wealth to the benefit of society.

Taxing income, the most common method of trying to redistribute wealth today, is ineffective in this regard. Though a large income over time can lead to wealth, they are not at all the same. It is not a disparity in earning power so much as hoarded wealth that defines inequality. A great hunter produced more meat than his fellows, but social rules then dictated how he was to share this with his community. The great provider accumulated wealth but rules in every society, like the west coast potlatch, then dictated how that wealth was to be transmuted into a community asset. If we want to produce a society in which we all feel ourselves to be an integral part of a shared whole, like the Athenian or the Florentine, we need a mechanism that induces accumulations of wealth to flow into community benefits. Incentives that favour hoarding can be replaced with those that favour the Cooperator.

There are persuasive reasons that this is appropriate as well as necessary. Most fortunes are actually based on public goods. Some have resulted from extracting and monetizing physical resources, monopolizing this process by claiming and enforcing a preferred access to the commons. The real owner of this wealth is society at large. Others were lucky enough to stand on the shoulders of a long line of innovators and inventors, some of them scientists in publicly funded research establishments, some of them unheralded tinkerers who, throughout history, have been responsible for many technological improvements. Such innovation is a broad process that ultimately rests on widespread education, the free movement and interaction of people and ideas, and robust legal protection for patents and contracts. These are all social goods. The real enabler of this wealth again is society as a whole. Still others owe the basis of their wealth to decades or centuries of exploiting people and despoiling environments leaving damaged commons and wounded people for the public to repair. This wealth is tarnished from the start and recompense to society is appropriate. Still others owe their wealth to gaining a monopoly, either by government fiat or by exploiting the unhealthy tendency of the system to concentrate power. That wealth too is tainted. Can any of this wealth be rightly said to belong to one person? Much of it is a public resource over which the rich are exercising stewardship.

Whether justified or not, however, huge concentrations of wealth do exist, another legacy of feudalism. This fact must be confronted if we value equality.

8. Taking Control

Redirecting this wealth into a community benefit in the face of reluctance would require serious inheritance and wealth taxes. However, since our goal is equality and a shared identity and not class warfare, we should put as much effort as possible into producing mechanisms to make the rich want to deploy their wealth to enhance the community. Philanthropy should be encouraged. The life of Bill Gates is an illustrative example. Some of his vast personal wealth was due to his personal qualities of drive and intelligence, but it was also crucial that he was at the right place, at the right time, with the right education and skills. A long line of technical and scientific innovations in hardware and software suddenly came together into a marketable form as the personal computer. By being the first to market a viable operating system, he was able to quickly gain a monopoly over a critical product, a monopoly he maintained and enhanced by using the heft of Microsoft to aggressively destroy potential competitors. These conditions—education, prior inventions, patent laws, court system and monopoly—made him fabulously wealthy.

However, during the second phase of his career, he turned to directing some of the wealth he had accumulated towards accomplishing social goals, such as the eradication of polio and the search for low carbon energy sources. He is not alone in choosing to dedicate accumulated wealth to accomplishing social goals. You can't take money with you when you die and there are many big challenges in the public sphere for people who like to solve problems. A few generations ago, the Ford Foundation, the Rockefeller Foundation, and the Carnegie Libraries were evidence that even the robber barons could feel the force of this logic. These natural stirrings towards philanthropy can be actively encouraged. Wealth and inheritance taxes can be used to nudge the wealthy in this direction. If people build libraries, schools or public buildings, fund education, science and innovation, commission public art or public performances, or provide transfers of wealth to directly eliminate poverty, all with their name prominently attached, they can strive for immortality as benefactors at the same time as the gulf between themselves and the majority are reduced. We should aim for a situation like ancient Athens where the meaning of wealth was an opportunity to enhance one's society. Transfers—wealth freely given to accomplish a stated goal—within communities as well as within the world, are an important aspect of active citizenship. We will come back to transfers later as they will be an important part of the structure based on cooperation that we develop later in the book.

We can reduce inequality. We can eliminate poverty. We can keep the wealth that capitalism concentrates from creating monopolies. Everyone can reflect their personal beliefs into the world through myriad acts of responsible consuming and active citizenship. We are not powerless.

The essence of the current deal, however, dictates that we accept the limited menu of options the machine presents to us. Every option on this menu has survived because it contributes to growth. Choices that would promote long term stability have been cut in the name of efficiency. Choices that would promote equality have been cut to give satisfaction to the Dominator. We must reject this powerlessness. We will now look at a structure that will allow us to do this.

CHAPTER NINE

Cooperation

Our ability to cooperate provides our alternative to dependence on the machine. Working together to solve problems has been an essential part of the human experience for hundreds of thousands of years. Being able to build on this strength, in place of basing our societies on domination and hoarding, gives us flexibility. Choosing to become the Cooperator confers a range of unexpected options.

Throughout tribal times, we lived in a family small enough to gather around a fire. A few such families making up a tribe provided a larger unit to approach larger needs. And when our tribe joined with others in the region for ceremonial gatherings, trade or war, we experienced the totality of our world. Every person outside this circle were "other." Cooperating in these three groups allowed people to accomplish a variety of tasks within a setting where they identified totally with their community.

Things changed drastically when the soldiers stepped above the group. Throughout feudal times, any empowered group was seen as a potential threat by the elites. Families, villages, towns, associations sharing an interest—all saw their scope whittled down as their prerogatives were taken over by those with the power to do so. They need to become empowered again.

Cooperation is not guaranteed just by putting a few people together. We are all driven to hoard, dominate and originate as strongly as we are driven to embed ourselves in a nurturing group. Because humans are a mix of these conflicting motivations, group experiences can produce many different storylines, sometimes with power delegated up an hierarchy from self-organized groups at the base and sometimes with power flowing down from Dominators at the top. Once the power relations are set, traditions and taboos will grow to maintain the status quo.

Whatever sort of organization emerges, it is enabled by the fact that humans have tamed themselves. We can work in large groups, coordinating the efforts of many to produce results far beyond the reach of the other primates. Put twenty chimpanzees on a bus and there will be carnage; twenty humans will sort themselves out with the help of a plan: ordered seats, a designated driver and a route schedule. Throughout feudal times, coordination was accomplished with whip and sword. Repeatedly, large groups of people were brought together to produce wonders such as the pyramids or abominations such as campaigns of war. However, though forced coordination can be effective, it is very different from cooperation, even though both rest on the ability of humans to work with each other.

For cooperation to occur, people must choose to join willingly in a shared effort. They will do so in proportion to the value they place on the group. This stance is natural to us—standing together to meet our needs undergirds many of our life experiences. Being a part of a nurturing group can be a source of meaning, comfort and satisfaction. Meeting our needs together breeds unity.

There is an interesting body of work by political scientist Elinor Ostrom that tells us much about the nature of successful empowered groups. She became interested in common pool resource problems—that type of public good where resources are finite and susceptible to overuse but it also can be difficult to exclude beneficiaries from extracting benefits—after reading "The Tragedy of the Commons" an essay by Garrett Hardin published in *Science* in 1968. He postulated that, if everyone has an equal right to a shared resource, it will eventually be destroyed because everyone will try to use more than a fair share. For example, if everyone has the right to graze animals on a common pasture, the pasture will surely be destroyed as everyone gradually adds one more cow or one more goat. Another emotive illustration he used was that of a lifeboat filled with more people than it can safely hold. He anticipated that it would surely be swamped as everyone tried to clamber on board. His conclusion was that cooperation in the face of scarcity was impossible, so all resources had to be privately owned, fenced for their own protection. He definitely saw people as principally motivated by the Dominator and the Hoarder.

This bothered her because she had experienced the successful cooperative allocation of a shared watercourse in a watershed in California. The arrangements she had experienced had worked for generations. As she looked for other counterexamples to Hardin's hypothesis, she found them everywhere from Maine lobster fisheries, to ancient waterways in Spain, to taxicabs in Nairobi. Groups cooperating to protect a limited resource were plentiful. But they were not universal—she also found examples of the disastrous situation that Hardin had anticipated. She dedicated her career to looking for the differences that would determine what made some groups successful while others failed. A lifetime spent following this question led to her becoming the first woman to be awarded the Nobel prize for economics in 2009.

She distilled her conclusions into eight rules for the creation of successful cooperating groups. I reproduce them here because they describe the setting we need in order to unleash our potential as Cooperators:

1. Define clear group boundaries.
2. Match rules governing use of common goods to local needs and conditions.
3. Ensure that those affected by the rules can participate in modifying the rules.
4. Make sure the rule-making rights of community members are respected by outside authorities.
5. Develop a system, carried out by community members, for monitoring members' behaviour.
6. Use graduated sanctions for rule violators.
7. Provide accessible, low-cost means for dispute resolution.
8. Build responsibility for governing the common resource in nested tiers from the lowest level up to the entire interconnected system.[1]

These guidelines just seem like good old fashioned common sense but it turns out that these precepts are often ignored or subverted when groups get together. Number 1 tells us to know who is in the group and who is not. Being a member must involve a commitment as opposed to just showing up. Number 2 tells us to agree about what we are trying to accomplish in this particular situation, to understand the resources we have available to make this happen and to make a plan that will match the two. Numbers 3 and 4 demand that the group be empowered to make and revise the rules governing group behaviour without being overruled or coerced by outside powers. Numbers 5, 6 and 7 demand that members of the group be held accountable in ways that encourage compliance rather than punishment and that the group have a procedure in place for dealing with disputes as they arise. Number 8 reminds us that we are part of a larger world and should engage with that world with the same respect with which we approach our group. It is not rocket science but having these precepts in view can help us avoid a lot of grief.

Effective empowered groups involve people dealing with people, so the size is not arbitrary. The essential interactions are conditioned by the physical features and abilities of human beings. As D'Arcy Wentworth Thompson stated elegantly in his classic text *On Growth and Form:*

> Everywhere, Nature works true to scale and everything has its proper size accordingly. Men and trees, birds and fishes, stars and star systems, have their appropriate dimensions and their more or less narrow range of absolute magnitudes. The scale of human observation lies within the narrow band of inches, feet or miles, all measured in terms drawn from ourselves or our doings.[2]

A human being interprets and influences her surroundings by walking amongst things, by touching and grasping things, by seeing things, by speaking to others, by hearing things, by smelling things, by tasting things. Our senses define the world we can know from experience. We can be aggregated into larger units using rules and force to bind us together but, as these groups grow beyond the human scale, no individual in them is able to sense directly what is happening or to effectively contribute to their character through her activities. These large units are defined somewhere else and controlled elsewhere. Their size makes them inherently disempowering.

The basic human group has always been the family—usually five to ten related individuals with a few adopted additions or extended family members blended in. Their various skills are sufficient to produce shelter, food and a safe space for raising the young. They meet regularly over a meal that they have cooperatively procured and prepared, an experience that personifies sharing. Our senses are engaged. We can look people in the eye over the fire or around the table and see them return our smile or evade our gaze. We can feel their body language, see how they listen, note how they engage with others, watch them show respect or disdain, appreciation or dislike, experience how they attempt to dominate or conciliate, help or hinder. Over many meals and other shared experiences, we gain a deep and intimate knowledge of who they are. Such a group can develop trust and cohesion. Families have always brought forth the acts of sacrifice and cooperation that made human societies work. Even as feudal violence defined the outside world, families continued to be a bastion of cooperation and support. The modern world has weakened this aspect of our lives by reducing the numbers in a family and scattering them around the globe.

The re-creation of this first human setting is where we must start. An empowered family, large enough to bring variety to our lives, will allow us to meet many needs that an individual cannot provide alone: meaning, emotional connection, financial support, child care, elder care, tending the ill, food gathering and preparation, and safety. Families do not guarantee trust, cooperation and safety as it is very possible for small groups to be dominated by a bully, but it is the setting where selfless cooperation is possible.

The family has never been sufficient. We are able to cooperate effectively in groups larger than this and we have needs that require the contributions of greater numbers to fulfil. A larger group in which everyone is known to each other through their daily interactions—a neighbourhood—provides a further set of opportunities. Historically, this group has numbered around one to two hundred people. There are many examples of groups that, when allowed the freedom to coalesce, iterate towards some number in this range.

This number has been defined by the anthropologist Robin Dunbar as "a cognitive limit to the number of people with whom one can maintain stable social relationships—relationships in which an individual knows who each person is and how each person relates to every other person."[3] While studying primate brains, he found a strong correlation between relative neocortex size and the size of the social grouping in the species. Extrapolating from the lower primates, he speculated that the human neocortex should allow us to function most comfortably in a social group with about 148 members. This has since come to be called the Dunbar number. This prediction has been shown to match well to groups that can function optimally in work and social settings. Groups larger than this sacrifice the ability to self-organize, interactions must increasingly be between people who know each other less well, and formal external direction becomes increasingly necessary. Groups smaller than this suffer from a lack of variety and sacrifice important powers of scale that diminish their effectiveness. Historically, this is our neighbourhood, our workplace unit, our military unit, the early American rural settlement.

This group cannot sit around a table with the intimate knowledge of others that is possible within the family.

Discussions need to be more structured. The model here is the town meeting in early America. At such a meeting, there are no strangers. Everyone carries a reputation which has been created over time through continuous contact. We all know who is wise, who fails to think things through, who is thoughtful, who is impulsive, who is generous, who is miserly, who is a bully, and who is driven by their own enthusiasms or grievances. This knowledge, gleaned by all in the course of their daily experiences, generates a natural structuring which allow fruitful discussions with a minimum of formality. The particular needs, the setting and the available resources shape these groups into entities with their own characteristics, preferences, expectations, strengths and weaknesses. Working together in groups of this size can focus the strengths of 150 individuals on shared goals. The neighbourhood is our second essential human-scale group.

We also need a larger setting—the area that we can walk through and see and hear and smell while interacting with more people than live in one neighbourhood. We feel comfortable in a world where we recognize everyone and know their place in the community, even when this knowledge is less personal and more functional than is possible in the neighbourhood. The size of such a unit, consistent with "ourselves or our doings," will vary depending on our doings. In tribal times, dispersed food sources made such large units next to impossible except on rare occasions. As farmers, the need for the village to be close to the fields imposed the same restriction. Only with urban settlements do these units become possible. Only when we could work together in these groups did we produce complex building and irrigation projects and scientific innovation.

In ancient Athens, a need for a ready phalanx demanded a citizenry of around 40,000 people, most of which were dispersed farmers supporting a denser town of around 5,000. In discussions, the words of all speakers were weighted by the power of their reputations. However, as numbers grew, the assemblies required more formal structures, more elected organizers and more rules.

The medieval city-states also numbered in the tens of thousands. However, the much smaller guilds provided the daily structure for the residents. These organizations fostered equality, providing meaning, variety, defence and mutual aid. Cities were strong because of these groups of Cooperators motivated by a felt connection to their fellows.

In the modern world before the car became king, this would be the rural town anchoring an agricultural region or the specific walkable area of a city. In these towns, interactions were frequent, character was visible and reputations were important. Everyone knew that they relied on everyone else. Patterns of mutual aid, such as barn raising, threshing bees, or volunteer fire brigades, generated enduring connections of trust. The adoption of the automobile weakened these towns fatally. The suburbs to which people moved had few shared public spaces, markets, nor nearby work to frame regular interactions. Needs were met distantly, making reliance on your neighbours unnecessary.

The family, the neighbourhood and the town have been the three human-scale groups throughout history. But people often still had the resources, knowledge and technical capacity to join together to address needs on a larger scale. In these groups, members were strangers to each other. Trust, knowledge and reputation were no longer part of the equation. The organizations they constructed had to be totally impersonal. Items like specialized hospitals can enrich our lives but they are impossible for human-scale groups to build and staff. Only when many towns share the effort will it be practical, but the planning, design, construction and operation must rely on statistics, measurements, rules, procedures and contracts. Such coordination can produce a useful tool. The nature of tools and ways to control them are questions to which we will return.

Cooperating is a powerful way to approach the world. Cooperative groups can become central to our lives again. The question now is: how do we go about it?

CHAPTER TEN

Organizing our Cooperation

∿→

The structure of our families has changed drastically since the end of the Second World War. Affluence, a recognition of women's aspirations to be more than homemakers, and the consequent opening of work opportunities to women have caused the size of the nuclear family to plummet. The affordable car, the creation of the suburbs, the globalization of industry and the introduction of cheap air travel have scattered extended families to distant corners of the globe, separating siblings and cleaving one generation from the next. It is now rare for extended families to have more than perfunctory contact.

Instead of living one's life in one place, embedded in a family in a neighbourhood in a town, every individual now is a part of many groups casually, usually superficially and often anonymously: one group of work colleagues, another group of strangers around us as we do our food shopping, a third group for social interactions, a fourth to share our exercising and so on. The rise of social media platforms has allowed people to substitute virtual communities for some of these human contacts. There, everyone can curate their posts, massaging the picture they display to the world, making this world untrustworthy by design. Instead of reputations becoming solidified over time, public personas are artefacts to be created and changed at will. These changes have created a world of isolated individuals. The family—half a dozen to ten people sitting around a table, gaining knowledge of others through interactions, eye contact and shared experiences—has largely ceased to exist. Neighbourhoods and towns have become collections of strangers.

The modern ideal is to go from door to car, often through an internal door to the garage, adjust the heat, tune in a soundscape of music or news, and move past surrounding houses in hermetic isolation. A significant part of the world of "ourselves or our doings" has become the inside of a car. We may meet neighbours for a summer barbecue or Christmas cocktails and nod hello occasionally, but there is little opportunity to gain knowledge or to form reputations. There are no empowered human-scale groups around us. We must get everything we need from strangers. There is no mutual reliance. There is no need for trust. The Cooperator has no place to act. We are left at the mercy of the machine.

It is possible to recreate human-scaled groups if we give them value. Consider this scenario, introduced here but built on later. A group of friends—a few retirees, a family with young children, a few singles, some related, most just friends—living in a city, decided to engage with the world as a family. They gathered over a glass of wine and a few pizzas (and sodas for the kids) and committed to do so. They found an appropriately large space and moved in. Similar families were coming to occupy spaces all around them and, as they shared stories, hopes, help and tools, trust grew and contact intensified. A neighbourhood came into being. Other neighbourhoods were coalescing nearby, filled with people with similar goals, so when tasks presented that could best be approached by neighbourhoods acting together, a meeting was called to explore these possibilities. The result was an empowered town structure.

They did not need permission from anyone to produce these human-scale groups. They had only to repurpose assets they already possessed. Of course there were challenges. Appropriate housing was rare at the start. But when they demanded a different type of housing and backed that demand with a willingness to spend, it started to appear. Cooperation worked and it was powerful.

No matter how effective the town became, it was soon clear that a few thousand people were not able to provide everything they would like to have in their lives. Towns joined with other towns to undertake bigger projects. People with whom they coordinated to do this were mostly strangers so this activity needed more formal structures. But they realized that these formal structures, or "tools," no matter how large, can never be allowed to operate outside of the responsible control of a human-scale group. I will look at building responsible tools here and then narrow my focus to the families themselves in the next chapter.

Tools

Tools are structures, such as hospitals, highways or universities, supported and used by so many people that it is impossible for everyone to join in a moderated discussion and have any meaningful say about their operation. Such decisions must be delegated to professionals. Our knowledge of these people comes from what they tell us rather than through long and varied contact. Their knowledge of us comes from surveys, reports and statistics and, as such, are seriously simplified. Tools are organized by goals, rules, procedures, inputs, flow charts for information and materials, and job descriptions. They are machines. When we demand their service and provide the resources, we cause them to act. Activities always have consequences and, as the user of the tool, we are responsible. Because of this, tools must be responsive to the wishes of their users. First, they need to have inputs that allow us to make clear, unambiguous and limited demands. Both we and they have to know what we are asking them to do. Allowing a tool to define its own scope of activity nurtures the Dominator. The tool would then have to be so transparent that everything flowing from our demands for service is open to scrutiny. And lastly, it would need a contract, regularly renewable, that specifies both the money to be provided and the service that will be expected. Design is important. It is not that difficult to empower the user. The design of such tools today has a diametrically opposite goal.

The resources we make available must be consciously provided for an agreed purpose. A general levy which guarantees support to a tool insulates it from our wishes. The professional organizers can then act as they will and we have no option but to accept. Our ability to initiate change by reacting to changes in our circumstance or our understanding has been removed. The tool, not the user, is in control. To be able to stand behind the chain of activity spawned by our demand for service, we must be capable of modifying or withdrawing support.

Many services today are controlled from afar only because distant agencies have usurped the power to do so. These services can be more effectively organized at the level of the town. This will bring the tool closer to the user. Locally designed projects meeting local needs with local resources can be controlled in a responsible way through local forums. We will feel, see, hear, smell and touch both our needs and the consequences of our demand for service. We will know the people deputized to carry out the task by character and reputation. Informal discussions can shape decisions. Our say is important. We feel the need, see the costs and must agree to provide the funding if we want the service. Changes can be dealt with in a timely manner at town assemblies when circumstances dictate.

Childcare, most levels of education, many medical services, elder care, welfare and local security are most effective when built on the relationships that one person already has with another. A tool that offers a credentialed professional on a schedule that serves their needs and with a continuity determined by their career aspirations badly mimics this human connection. Other tasks that affect only our local space such as parks, parking, local roads or garbage removal are also best designed and controlled totally within an area of people's doings. Coordination with other localities doing similar tasks can easily be arranged if the need arises but the choice to coordinate or not would be made locally. Within their town, people can meet needs responsibly because they themselves will experience the effects, shoulder the costs, and enjoy the benefits. Irresponsible activity, if it arises, can be noted, named, countered and stopped.

Decentralization of services to the local is generally opposed in the name of standardization with its inherent promise to redistribute wealth invisibly from richer to poorer areas—ideally all will get the same service while the tax contributions will differ. But these transfers of wealth are only necessary because of persisting inequalities. Constructing centralized disempowering tools to address this inequality locks both disempowerment and inequality into a structure. Inequality should be faced directly as a problem instead of accepting it as a design feature for the whole of society. The barriers that sustain poverty can be attacked with jobs, mentoring, improved wages, better education, help with childcare, improved infrastructure, better access to health services and other activities that require thought, time, effort and commitment as well as money. We will come back to both transfers and equality again later in the chapter.

However, even after many services have become locally-based, there will still be a need for some tools, such as a specialist hospital, which can only be addressed by large coalitions serving a huge area. If possible, these could be structured on a fee-for-service basis, either paid at each use or as a retainer with a time period that allowed the tool an adequate, though not indefinite, planning horizon. Individuals should always be able to choose to purchase the service, search for an alternative or go without. This retainer could be paid by individuals or organized through the town if there was a recognized social good for everyone to have access to the service. If some in the town disagreed with such universal access, then compromises must be sought to allow individuals control over their degree of participation. Perhaps some people could reduce their contributions while accepting limitations on their use of the service. Perhaps they could pay in labour or have their contribution forgiven if lack of resources was the issue. A discussion where everyone is known by character and reputation favours the crafting of compromise that will work for both the success of the tool and the empowerment of the people involved.

When the service being delivered is a general public good like a highway or an air force, fee for service may be more difficult to arrange while it is still essential that control over funding rest with the individual. There must always be a visible user for every tool activity. That possibility of feedback is what allows us to solve problems before they become chronic.

It would be useful to receive a regular statement showing our expected contributions to each of these tools with the benefit expected and the contracted period. Evaluation of their performance by an independent agency would be available, just as was the case for firms in the marketplace. At specified intervals, we could choose to reduce or increase the level of service we desired or give notice of our intention to opt out. If we note things that are troubling, we can request changes backed by the real possibility of discontinuing or downgrading our engagement. For services we value, we would come to understand why they need our money—we would have to assure they had adequate resources to do their job properly. It would be like dealing with tax returns today augmented by feedback and an ability to take action based on what we see.

For some tools that are costly to build and have a different set of costs to operate, such as the specialist hospital, splitting the capital costs from the operational costs would be helpful. If the capital costs are guaranteed by a few towns, then the facility can be built. Every town does not have to be involved for the project to proceed, just enough of them to cover the capital costs. Towns deciding not to join could have their reticence respected without making the project impossible. Once the facility is running, then the operating costs could be covered by a fee for service. The service could still be offered to people who had opted out of the original capital contribution but for a larger fee that represented the capital they were using along with the costs of operation. Dissenters could decide to purchase a capital share later. There are myriad ways to build tools which allow responsible control. How this is done is only a design problem. Currently, most public agencies are guaranteed both resources and unquestioned control. As a consequence, they are strongly resistant to change when their usefulness decreases or giving good service becomes inconvenient.

If so many people opt out of a project that its finances are jeopardized, then the tool should be either redesigned or shut down. When people want to produce their own alternatives, we leave the door open to new ideas. For example, if a sewage system sees some people opt out because they believe that composting their waste is more responsible, then their refusal to support a treatment plant illuminates this other option. They may find that it is too much work and later buy into the community project. On the other hand, they may develop a cheaper, more responsible option for dealing with waste that we can all adopt in the future.

The Hoarder is always with us. Even with good design, the free rider problem—people wanting a benefit while looking for ways not to pay for it—cannot be entirely eliminated. There will always be some services, such as the beautification of a town square, where it will be impossible to limit benefits only to those who choose to subscribe. To avoid negating personal responsibility by forcing contributions, we must try to find consensus. First, instead of just rounding up the votes necessary for a majority we must try to understand the issues that prevent dissenters from embracing the project. If lack of funds is the problem, then compromises allowing other forms of contribution could bring everyone together. If so many people are in this boat that the project is threatened, then perhaps we should not be beautifying city squares when people are having to choose between this and basic needs like food. Reluctance is a form of feedback that should be valued and examined in detail. If the difficulty lies with specific details of the project, then compromises that incorporate differing views into the design should be sought. A project that reflects the community even if it is not exactly what any particular person would see as ideal may be what we need at this point.

In forums where people are known to each other in many aspects of their lives, differences can be uncovered and compromises sought. But there may still be times when, with the best of efforts and intentions, a compromise will be impossible. If the dissenters are a small minority standing against the wishes of the majority, then we can respectfully ask them to recognize the general will and let a desire for community harmony outweigh their objections. If this fails, then we could, in the same spirit of compromise and desire for cohesion, forgive their financial contribution, asking them to contribute nothing but their acceptance of a project in their midst with which they disagree. If their refusal rests on such a deeply held conviction that they still cannot budge, then we must consider abandoning the project. However, if we are convinced that their refusal is motivated by either a desire to get a free ride or to impose their individual control in place of consensus, then the issue facing us is the place of those individuals in the community. If either motivation is deemed consistent with their reputation, then they may be a bad fit for the community. Community membership carries a responsibility to listen, to help, to contribute and to value the group. It is not claimed by just showing up. Consensus will generally be within reach in committed communities. The free rider problem should be rare in healthy human-scale groups but, when it does arise, it and not the particular project, is the problem that must be addressed.

Transfers and Equality

Responsible control of our tools demands that we be able to contribute or withhold the resources that fund every good or service. But, if we say no to taxes that promise to transfer resources to the more needy, some people will lose supports on which they have come to rely.

Communities, as well as individuals, differ in wealth, a legacy of centuries of feudalism, exploitation and colonialism. These systemic inequalities have generated pockets of poverty within our towns and cities, they have led to the existence of relatively poorer cities within our region, and they cause areas elsewhere in the world to live close to starvation. It is essential to a world of peace that these inequalities be eradicated. When rich and poor must coexist, there is fear, envy, anger and violence. Attacking inequality reduces these evils at their root. We can't change history, but we don't have to accept its legacy.

Transfers are responsible only when donors freely decide to provide the resources. When we are convinced that our lives will be improved by reducing inequality, then we will signal our willingness to support ways to do this. This expressed demand will elicit proposals that target barriers instead of just offering to redistribute money. They would present a specific goal, a definite timeline, a method to assess whether their efforts caused inequality to decrease and a degree of transparency that would allow evaluation. If a scheme was seen not to be effective, then we would be able to reallocate our support to other approaches, learning from this failure but continuing to search for results. Two factors lead me to believe that transfers based on free choices would actually increase from their present hidden levels.

The first is self-interest. Neighbourhoods are weakened when they contain a family unable to meet their basic needs, let alone contribute to community projects. Neighbourhoods are strengthened when all families are standing on an equal footing. If allowed to persist, this inequality would darken our daily interactions. However, with the knowledge gained from regular contact, neighbours would be best positioned to pinpoint the type of help that would make a lasting difference.

If poverty was caused by the lack of an appropriate job, then the other families could call their contacts, cosign loans to startup businesses, help cover the costs of appropriate education, share skills to mentor someone in a trade, or even share their own work. The transfer would manifest as an appropriate job. Success would be easy to measure. The unequal situation would disappear. The cost here is more in effort than in money. If poverty flows from physical limitations, then appropriate supports could be put in place while a job tailored to the abilities of the individual could be sought. If the barrier is addiction or mental health issues, then the neighbourhood is the proper place to produce a harm reduction strategy because of the constant presence of an enveloping team of supporters. These transfers all involve identifying and removing the barriers to community engagement. Monetary transfers would be only a small part of the solution. The benefits to the recipient family of being on an equal footing with their neighbours would be huge, but the benefits to the neighbourhood of having everyone standing together as equals would also be immense. These transfers would be a one-time effort targeted at specific barriers, a response different in kind than establishing an agency guaranteeing to support an unequal situation.

At the level of the town, the same argument would apply. Hunger and need on our streets, in our parks and around our town shames us all. When this poverty forces people to turn to crime, everyone is less safe. Walking past obvious need quashes our joy. Again, our contact with this chronic poverty should inform us about the specific barriers and help point us towards appropriate solutions. If jobs are needed, then they should be found. If changes to housing stock, schools or parks are required, then donations as well as work parties could help make these changes real. If a lack of a town service such as public transport is a contributing factor, then that can be corrected. The scenario is the same. Identify the barriers. Make a plan to which people will agree to contribute. Eliminate the barrier once and for all.

In the opposite situation where inequality manifests as one neighbourhood being significantly more wealthy than the rest, cooperation will also be impaired. Different goals, different fears and differences in resources will lead people to activities specific to their class instead of embracing cooperative ventures that span the whole community. As goals fail to align, there will be conflict where they meet. Their relations will call more on the Dominator than on the Cooperator. This dynamic doomed Athens and the medieval cities. It is a danger, but it does not have to be.

As true security comes to be seen by all as resting on being part of a nurturing community, then expenditures that enhance the whole community will become more common. Philanthropy will become an admired stance for the wealthy. Others can encourage this by granting esteem. Perhaps, if it is deemed necessary, a tax code that nudges people in this direction will be necessary. The goal is that the wealthy should want to take this road in the belief that a stronger community of which they are an integral part is more precious than their gold. It happened in Athens; it happened in Florence; it can happen again. Whether the process begins with the wealthy wanting to be a more included part of the community or whether it starts with the rest of the town questioning positions of privilege, attitudes that value community and that devalue displays of wealth will lead to change.

We share concerns with other towns in our city and other cities in our region. There is much to gain by cooperating in the creation of tools such as highways, flood control systems or specialized hospitals. Inequalities here will be a source of friction as richer areas push for services that poorer areas will not be able to afford. However, with transfers between towns, people no longer know others by reputation, so these transfers must be organized through a tool. The approach, however, would be the same: engage with the intended recipients to identify specific barriers and then generate targeted programs to reduce their effect. The barriers could once again be lack of jobs, or jobs that do not pay a living wage. Improving the wages flowing into a town will allow people to deploy their own resources to solve their own problems. It is amazing how easily poverty can be reduced with the provision of a reliable cash flow. Help in establishing new businesses in the town may be required. Other barriers could include poor educational facilities, inadequate health services, lack of access to transportation, or substandard housing stock that is controlled elsewhere—these can all be changed using effort and a little bit of money. The tool would engage, plan, seek support, organize activity and measure results. In some cases, a guaranteed income for a defined period could allow a town to demolish their own barriers. Tools can work. Distant inequalities can be banished.

This same argument holds true over even larger areas. It benefits everyone when inequality is reduced between provinces, between states and even between continents. Friction, resentment, envy, rebelliousness and fear can be lessened at every scale. As distance increases, the tools must change, the identified barriers will change and the programs will be different, but the methodology would be the same. Identify specific barriers. Create a plan. Recruit support. Measure results. Dissolve the tool when the goals are met. Our lives will improve as we chip away at inequality on every scale. As we feel the benefits, we will look for opportunities to remove more barriers. It is a very positive feedback loop. Transfers will not dry up. They should increase as inequality decreases until that day when they truly are no longer necessary.

The second factor prompting transfers at all levels is responsibility. The reason we demand the right to direct our resources is to assure that all activity done in our name is done in a responsible manner. We must be prepared to stand behind every action we initiate, answering for harm and making wrongs right. Inequality, in every case at every level, exists for a reason. It may be that a city, a province, a region or a continent is poor because a Dominator appropriated their resources, thereby impoverishing the area. It may be that invaders crushed their ancestors in war, enslaved or dispossessed them, and their descendants have never recovered. Our shared history has created the structures we have inherited. If we rebuild our world by redeploying our wealth while ignoring that others cannot, we reinforce old injustices. It is possible to replace the exploitation embedded in the structure with equality. Recognizing that, we have no other responsible choice.

Equality over great distances produces real gains. As international inequality decreases, tensions between areas will be replaced with bonds of travel, trade and shared projects. That lets us dream of the day when we eliminate war. The chain that leads from inequality to resentment to rebelliousness to fear to the creation of armies to using them can be broken only when we eliminate the original cause—the inequality. Acts connecting distant regions through projects designed to lessen inequality generate a different chain of consequences. These acts of cooperation nudge us towards unity rather than enmity. Empowerment will produce a world that reflects our beliefs. Responsibility demands that war have no part in it. This demands that we unrelentingly attack inequality.

Unleashing Active Citizenship

Government must be greater than just a collection of tools with the legal right to tax us to pay for them. We need serious deliberative bodies of various sizes to explore ideas about the nature, the possibilities, the risks and the future of the human adventure. These councils would take up the task of examining our shared needs and fears. We are an eight-billion-person family. We need a structure that will bring us together.

Currently, true to their origins in our feudal past, governments disempower individuals while separating them into various interest groups. Transparency is actively opposed. Information is twisted to win elections and solidify control. Facts are used selectively as political weapons instead of being treasured as a basis for understanding. Our governments are machines of authority. Our relationship to them is entirely passive.

We need a structure that accurately and transparently distills the beliefs of many into a consensus. An attempt at such a structure was made in ancient Athens, and again in the medieval city-states. Both had some success. However, both ultimately failed as people clung to their inherited inequalities. This corrupted their governments, returning them to arenas for amassing power. A community of empowered responsible individuals can try this again. We have the benefit of their examples. We also have a range of technical communication machinery that was unavailable to them. We have a clear vision of what we need this structure to accomplish. Global active citizenship is a possibility.

As in everything else, transparency is crucial. In every forum, there must be independent observers bringing information to us about the ideas under discussion with as much abbreviation as possible coupled with as much detail as we need. This dissemination of information should never be left to the forum itself—they are human and will surely fall prey to the promptings of the Dominator. Representatives must discuss, deliberate and propose; independent observers, funded by and answerable to us, must be our eyes and ears as they describe and report.

The real challenge will be devising the structures to send information the other way. There must be channels by which people can reflect their beliefs into this world. This points us in two directions. First, our representatives should be extensions of smaller groups. Our city representative should be a member of the town assembly, a voice for a consensus arrived at in a human-scale discussion. She will bring discussions at the city level back to her town. Regional councils should be made up of representatives from city assemblies and so on. The contact between the people and progressively larger levels will become weaker but efforts to keep this conduit open and honest can help to ground most bodies in their roots.

Careful use of referenda through our ubiquitous devices can aid this process. In this, the framing of questions is crucial but it will be more likely to be successful if the goal is gaining understanding rather than shoring up some political stance as is the case today. Such gathering of opinion can pinpoint differences between regions, interests or philosophies and uncover the need for compromise. This sends us down the same road we trod in seeking consensus in human-scale groups. Altering a project to remove objectionable aspects may work. Adding options to allow people to use tools differently may be necessary. Ultimately, respectfully asking a dissident group to accept a project despite disagreement could allow it to advance without rancour. Consensus seeks that place where every voice is heard and every belief can find a place. Referenda can be a useful tool in helping civic bodies arrive at the consensus which accurately reflects the beliefs of the people.

When we are considering the potential of civic bodies, we should be careful about judging possible structures in light of our own world where apathy, induced by generations of powerlessness, is the logical choice and where control, not consensus, is the goal. In a world where thoughts, ideas, understanding and feedback matter, where we can truly believe that our efforts can create an embodiment of our shared hopes and fears, people will participate. They will look for compromises. They will engage. Citizenship will become active.

Beliefs put into action can transform the marketplace, our tools and our civic structures, so we must be very clear about what we value. If we truly value community, our every action will cause it to strengthen. If we truly value equality, our actions will generate an equal world. If we believe that everyone deserves respect, we will live in a respectful world. If we believe that citizenship must be active, we will not be controlled. If we believe that no one should be hungry or poor, we can banish war.

On the other hand, if we continue to believe that the lowest price is the only goal worth pursuing, then we will live in an exploitive world. If we believe that we can only prosper if someone else suffers, then we will live in a world with desperation, fear, envy, violence, crime and suffering. If we believe that one vote every four years gives an elite the right to rule over us, then we will be controlled. The choices are clear and they are totally ours.

We can use these precepts to describe a new world. We recognize the importance of the three human-scale groups and have seen how they could function. We know how to create controllable tools. We know that it is possible to banish inequality with transfers if we are skillful and persistent. And we can aspire to build civic bodies that will express and focus the will of the people. When we put these elements together, we get exciting possibilities.

This new world will not emerge from current historical processes as we stand by and watch. Anything spawned by the present system will be sure to have growth and competition at its core. There is no solution to the problem of growth here.

Nor will change issue from a movement that asks everyone to adhere to a specific blueprint that enforces cooperation—the communist utopia. This kind of predetermined future denies the responsible choices that are necessary to reflect our personal beliefs into the world. It reduces people to just those aspects of themselves that can be plugged into the blueprint. It attempts to create an ideal machine, one that differs from our present leviathan, but a machine nonetheless. It denies the dynamism of community every bit as effectively as the machine we already have. The makers of the blueprint and those charged with enforcing it are more powerful than others. Such an approach must fail, regardless of the beauty of the blueprint.

Rather, change must be the result of myriad free choices taken in the belief that they will make us, our children and future generations happy and secure. The acts of leaders will be irrelevant; those of billions of individuals will be crucial. As we each become aware of the flaw at the heart of our survival strategy, we will have no option but to act. Each look at our grandchildren will demand it. Changing our personal world will be our choice. The larger picture will follow.

We must each ask ourselves if the deal still works for us. Will we be better fed, more secure and happier if we continue to support the status quo or is it necessary to change our priorities so that we give value to cooperative families and neighbourhoods? Can we act to create a life where our beliefs matter and our choices reflect these beliefs into the world? What do we want to be next year? In ten years? In fifty? What kind of world do we want to leave to our great grandchildren? Engaging with these questions is the start.

Our actions as consumers, voters, workers, parents, friends and neighbours continually recreate the world. Our beliefs give direction to this change. The problems against which we rail, demanding that someone else provide a fix, are simply a summation of choices made by individuals. These beliefs, choices and acts can change. A script is part of the deal, but we choose to follow that script. We do not have to.

Humanity has endured enough inequality, poverty, war and violence over the last ten thousand years to learn that a society organized by the Dominator is flawed. The current version of this world produces tons of stuff at a cost of tons of suffering and a creeping destruction of the commons. We cannot afford not to act.

We have both sufficient knowledge and sufficient wealth to create any possible future. But reluctance to abandon a survival strategy is an ingrained part of our nature. Change must be gradual, not abrupt, growing within the present in an evolutionary way. That route is open to us. We are both Dominator and Cooperator and can make small choices that favour one over the other. We have options. We need only an example of how this can be done. And that is where we go in the next chapter.

Chapter Eleven

Eve

〜➤

Consider Eve, an idealistic young woman who was in considerable despair about the world she was leaving to future generations. She had foregone motherhood in a mistaken belief that the future was no place for children. She saw that the course the world was travelling would inevitably end in tears. Sea and land pollution, regular crises generated by a changing climate, habitat destruction and species extinctions were all seen as the price of her comfortable life. She knew in her heart that the present course was not sustainable. Her helplessness in the face of looming disasters sapped the savour from her life.

She had become isolated from family and friends as she followed her career around the country, joining a new group of coworkers every few years. The pleasures she used to take in her sleek apartment and from the purchase of a nice car or fashionable clothes had disappeared as she began to connect her casual consumption with the problems around her. Her full closets and groaning shelves became a constant silent rebuke. She desperately wanted to become part of a solution instead of seeing herself as a part of a problem. Her daily activities no longer provided either meaning or assurance. She was desperate to make a change.

She did what she could by herself. She attacked her carbon footprint by reducing the number of her purchases and by consciously directing the ones she retained toward producers that she believed should thrive. She sold her car and bought a bicycle, foregoing some activities in the name of simplicity. She lowered her thermostat. She tried to reduce the volume of her waste, waging a private war against plastic. She recycled assiduously. She petitioned government agencies to take strong action against environmental destruction and got a form letter back telling her that the professionals had it in hand—be patient and vote for their party in two years' time. She joined lobbying groups and they put her on a list that cluttered her email with regular requests for donations. She changed careers, becoming a social worker to try to mend some of the harm she saw around her. But those she helped today were back at her desk tomorrow, every bit as hungry and needful as before. It seemed to her that the poor stayed poor, the rich built higher fences, and her job was to apply the Band-Aids. She was not healing the world as she had hoped. Her efforts seemed somehow divorced from the problems. She was always arriving after the damage was done.

She wanted to know why things were like this. In her experience, most people were intelligent, competent and caring individuals. As a species, we were wealthy, resourceful and had proven capable of great things. If we could send men to the moon and bring them back, why couldn't we eliminate hunger within her city? She wrestled with the various answers offered in lectures, documentaries, books and conversations. As her questions became more pointed, relevant books, people and information seemed to be drawn into her path, each offering another brick for the wall of understanding she was constructing. Gradually a picture emerged that made sense to her. She caught a glimpse of the Dominator lurking in the rules of the machine.

As her understanding grew, she felt impelled to try to incorporate it into her life. Talking to friends, acquaintances and relatives, she found others who shared her concerns. Their perspectives helped her refine her picture. Occasionally she and her friends encountered ideas that were hopeful, examples of small successes or inspiring individuals. These usually exemplified human acts of selflessness and cooperation. Wanting to bring that feeling of hope into their lives, wanting to combat the isolation they all felt, and wanting to simplify their lives, a group of friends, including Eve, decided to join together to create a setting where their cooperation would be encouraged. They decided to form an intentional community—a family.

Intentional communities are certainly not a new phenomenon, and in their reading they had learned of many. From tribes to communes, to religious groups formed around charismatic leaders, to communities cleaving to each other in the face of existential threats, cooperating groups have always been able to offer opportunities not available to isolated individuals. In the most successful of these groups, individuals came to identify so strongly with the group that it became a virtual extension of the self. But it was also clear to them that not all groups endured—just clumping people together does not add to anyone's life. Casual groupings would generally dissipate after a short honeymoon as members drifted on to other enthusiasms. However, groups that were committed could endure for decades or centuries, providing islands of cooperative activity within the competitive world that had spawned them. The group of friends resolved to become a group that would last.

One of the bricks in their wall of understanding was Elinor Ostrom's principles for constructing groups that could cooperate effectively. They addressed each of Ostrom's points in turn. Boundaries around their group were established by demanding of each other a commitment that would make clear to everyone who was in and who was not. They left no iffy middle ground for people to slip in and out as the spirit moved them. They set out ways for respectfully voiding this commitment if someone needed to leave and ways that outsiders could petition to join, but it was agreed that neither of these courses of action could be undertaken casually or unilaterally. The membership of their group was made clear, both to them and to the rest of the world.

As they shared their visions of what they each wanted the group to be for them, common goals began to emerge. They agreed to share living arrangements, child rearing, elder support, care in times of sickness, food preparation and financial support. With those goals as direction, the path forward became clearer. They turned to Elinor Ostrom's next points about the importance of defining methods of arbitration and mediation in advance so they would be able to deal respectfully with differences. They set out norms for discussion and action that ensured that all would be heard, no one would be able to dominate, all would share the tasks, and each would submit to group adjudication of problematic quarrels. This became their family structure—defined membership, agreed goals, expected norms of behaviour and an established way to handle disputes. Two of the original band decided that commitment to this structure was not for them. They left. But others, attracted by the project, petitioned to join and were accepted into the group. Sharing their hopes and committing to each other strengthened the connections between them. As their family took form, each felt themselves to be part of something larger than themselves. Eve felt less alone. She now had partners in her quest to change her life and the world. She had direction. Her life was no longer totally meaningless.

To support each other as they wanted, they needed to be able to live together. They went looking for an appropriate space—a unit that had a large bedroom for everyone, a kitchen large enough to prepare and share meals for about a dozen, and a shared living room large enough to be comfortable when all were at home. Not surprisingly, there was nothing on the market that immediately met their needs. They decided they would have to construct a space tailored just for them.

There were fourteen people in their group. Four, of which Eve was one, were singles who had been renting their own apartments. There was a husband-wife pair with an eight-year-old and a twelve-year-old who had owned their own home while still paying a mortgage. There was a single parent with a three-year-old, who had been renting a small house. There were two retirees. One was a widow who owned a large house while the other, a divorced man who was the father of the single parent, had been living alone in a condo. Both retirees had already paid off their mortgages. There was also a couple, living together and sharing rent, but whose level of commitment to each other was still uncertain—they existed someplace between roommates and a couple. So, in total, their family comprised three children, two retirees, and nine working-age adults. Eight had been paying rent. There was one mortgage and two properties with clear title. One of the adults was unemployed and seeking work, while eight were carrying full-time jobs. Five of them saw work mainly as a necessity to obtain a pay cheque; the other three had fulfilling jobs through which they saw themselves contributing to the world. The two retirees had pension income.

As they began their search for a home, they found that they were not alone. There were other families forming and looking for housing appropriate to a dozen or so people. The idea of facing the world by cooperating in families had recently become part of the global discussion and this was giving direction to a widespread feeling of discontent. Though each family was unique—a reflection of the people in it—they were all looking for a space of about eight to ten thousand square feet that was open to serious renovation. These groups found each other, began to talk and saw merit in joining forces. Five families combined their resources and bought a solid old six-story apartment building, with each floor currently having ten thousand square feet divided into nine units of just over one thousand square feet each.

Our group had resources. Those who had previously been paying rent or making a mortgage payment could direct that money towards their new home. Those who had owned outright sold their properties, so they had cash. With eight incomes and two pensions, they had considerable cash flow.

The five families commissioned an engineer to identify the structural aspects of the building. They then engaged an architect to help them translate their particular hopes into a plan for each floor. A portion of the ground floor, which had been purchased jointly but not occupied by any family, was fitted out as a workshop to backstop the renovations. Another area became a temporary storage area for building materials that they could purchase jointly in bulk. They brought the tools that the five families already owned to the workshop; others were purchased to fit it out as a common resource. The renovations began.

The removal of the non-bearing walls became a revelation of space as hallways and the walls that had existed solely to create many small rooms disappeared. The size of their new home suddenly revealed itself. Since the building was structurally sound, most of the work—framing new non-bearing walls, drywalling, laying floors, painting—were things they could undertake or help with themselves. With more than a dozen members in each family, and a little help from their friends, the new units came together quickly. The nine stoves, nine fridges, nine washers, and nine dryers which had come with each floor were sold, replaced by a few robust, industrially-efficient models more suited to serving a group of this size.

Each person was allocated about 250 square feet for personal space. The young family decided to configure their 1,000 square foot allocation into three relatively small bedrooms and a large play/family room/private activity space. The couple wanted two rooms, each with doors into the common area and a connecting door between them—still navigating that space between roommate and married. The single mother with the three-year-old centred their 500 square feet around a play area that could remain scattered with toys, and two small bedrooms. Everybody else laid out their fifteen by sixteen foot space, ample for at least a desk and easy chair as well as a bed. Eve wanted her space to be ten feet by twenty-five, to provide two distinct functional areas—bed and lounge—that she could separate when needed with a shoji screen. These desires quickly became real when the walls went up. The fourteen personal units of 250 square feet took up 3,500 square feet. The other 6,500 square feet became kitchen, lounge, activity rooms, and service areas such as laundry, bathrooms, stairwells and elevator. There was space for these rooms to be more than ample. They chose the best of the furniture they already possessed from their nine homes, sold the rest, rented a truck to gather their belongings from around the city and moved in.

They shared chores. Everyone was the designated cook one night every two weeks. The children needed help on their nights but that was shared around so that it did not always fall to the parents. As the kids gained in competence, the adult who was present with them became more of an aide, being there, answering the occasional question, but not having to be particularly active. The children were expected to be responsible agents. Preparing a meal for fourteen involved some larger tasks than they had been used to—peeling potatoes for fourteen definitely took longer, but putting fourteen salmon steaks in the oven or setting the table with a larger stack of plates took not much longer than it had done for one or two. The night that they were cooking required them to start early and dedicate half of an afternoon. However, on the next thirteen nights, they could appear at the supper hour and sit down to a prepared meal. Because of the rarity of cooking, they started to look forward to their turn as a chance to place a really good meal before the family. People lingered over these shared dinners and the conversations roamed far and wide. They knew each other a bit better each day. Two people were assigned to dinner cleanup and dishes, another of the chores, meaning that this too fell to each person only twice every two weeks.

The cleaning of the common areas was divided into fourteen tasks—clean one bathroom, wash one floor, dust one section—each small, each rotated through the group week by week. If some chores were unwanted, such as cleaning a particular bathroom, this task could be done in the knowledge that it would not fall to you for another fourteen weeks. As was the case with cooking, rarity turned what had once been a chore into a gift for our friends. The sharing of tasks seemed almost to make them disappear.

Pressure on parents was reduced because there were always people around to keep an eye and an ear on the kids. The kids, for their part, developed relationships with eleven different adults, each sharing different skills, each with different ways of relating. If the parents were out, the youngsters could stay home, with no need of a babysitter and no disruption to their routine, with adults nearby as a matter of course.

Four of the family were working remotely. The numbers were similar for each of the other families in the building. Altogether, seventeen people in the building were working from home. They decided to rent shared office space a short walk away, allowing them to share support services, experience a supportive work group, and move that function out of the home. Two other people in the family were already working together, doing a similar job in the same business. They decided that they could share their work with the friend who was unemployed, turning two jobs into three with shorter hours. They each did take home a bit less money but they appreciated the time they gained. And the drop in their take home pay was actually less than one third since the hours that were foregone were hours that had been taxed at the highest marginal rate. Sharing ten working days per week between three people meant they each worked three days a week for two weeks and four days on the third. They found the change in work-life balance transformed their lives: after four days away, they were eager to return to work; after three days at work, they were ready to get back to the projects their new leisure allowed them to take on. They no longer had to cram their enthusiasms into a weekend that had always seemed too short. The new cooperative arrangement had major effects on everyone's work life. We will expand on the nature of work in a cooperative world in the next chapter.

The group found that they needed fewer cars. Only two of their cars were now needed for commuting as four of the nine workers walked to the nearby office for their remote work, one was well served by public transport, and the three that were sharing work could share a car because they travelled to the same workplace. Only one of them needed a car for access to dispersed clients. Beyond these commuting needs, two shared cars were found to be more than enough. A sheet tacked up on the kitchen bulletin board allowed people to reserve one of these cars when it was needed. Five cars were sold. Five insurance bills ceased to arrive; five servicing and repair responsibilities disappeared; five parking spots became empty.

Living together was making their lives less expensive all the time. Utility costs per person plummeted. One share of the mortgage they shared was less than any of them had paid in rent individually. They needed to buy fewer appliances and furnishings. They saved on childcare and food. Car expenses had been more than halved. As expenses dropped, they could look at reducing the hours they all had to work.

Other people were attracted to the unfolding adventure and gravitated to Eve's part of the city hoping to replicate their arrangement. Across the street a group of ten townhouses and three four-story low rises had been renovated and occupied by families of similar size. The two families in the townhouses had been able to open some of the connecting walls to make adequate, though sprawling, homes. Each low-rise housed two families. These eight families welcomed the ideas, advice and volunteered labour of their more established neighbours as they set themselves up. There was much mingling. Friendships and reputations grew through the shared work. These new families bought into the common area, letting them share the workshop. A neighbourhood had formed.

The space between these buildings was busy with comings and goings. Everyone thought this area could be improved so an ad hoc committee came up with a design incorporating paths, benches, plantings, gathering places and playgrounds. The parking lots shrank. Work bees turned the plan into reality.

As the outdoor and indoor shared space was more intensively used, chance meetings became more frequent. Everyone in the neighbourhood became well known to everyone else. Children found playmates within shouting distance. Neighbours engaged in conversation, swapped books, and shared sewing, guitar and computer skills. When it became clear that no one in the neighbourhood had much in the way of practical medical knowledge, a desired resource, two people volunteered to take an industrial first aid course. When one person unexpectedly lost his job, the news spread. The next night, a neighbour dropped in to mention that he had a friend whose firm was hiring—he would make a call and vouch for him. These people became a functioning neighbourhood, helping each other as they shared their lives.

They found that even twenty-five cars in readiness for thirteen families was more than enough. With a reservation system on their phones replacing the sign-up sheets, they found that ten were sufficient. More savings. More land reclaimed from parking. Fewer cars cluttering the streets.

There were savings at every turn. They needed less money to live at the same time as they were valuing time away from their work to be in their homes and neighbourhood. Everyone took a few less shifts or a few less projects. Soon, a three-day work week was the norm rather than the exception. There were more people in the neighbourhood at all hours. Acts of caring surrounded them more and more. Eve felt motivated, involved, secure and satisfied. She knew who she was and much of this sense of meaning came from her community. Looking around her, she started to feel hope for the future.

If this had been as far as things went, these 150 or so lives would have been changed radically, everyone gaining time and balance in their lives, everyone being able to reduce their consumption, everyone being able to reduce their expenditures, everyone losing their isolation and gaining an enveloping community. It would not have addressed larger questions in the world, but their lives would be richer. However, in Eve's story, it went further.

Similar families and neighbourhoods were coming into being elsewhere. To the east, a group of three twelve-story buildings had become three neighbourhoods. To the north, between Eve's neighbourhood and a large park, a dense area of detached homes had created three neighbourhoods spread over what had been six city blocks. To the west, an area of mixed building types similar to theirs had resulted in five more neighbourhoods. To the south, a major thoroughfare one block over effectively cut them off from casual interactions with others. Together, these twelve neighbourhoods made up a connected area containing nearly two thousand people between the highway and the park. This area was about four blocks by four, easily walkable. Everyone who lived here shared similar challenges, had undertaken similar tasks and were experiencing the same benefits. They had much in common. Their lives overlapped in many places as people moved through this space.

A desire to improve the ability to walk between neighbourhoods spurred the first town meeting. A few moderated open houses allowed a committee to come up with a design and they got to work transforming the space. No longer needing most of the parking spots gave the town new areas that could become green in various ways. People walked the new paths, sat under the new trees, ran in the new parks, and used the new benches with an expectation of connecting because these others in the shared space were not strangers. They were aspects of "ourselves or our doings." Friendships grew. News and gossip travelled. A web of connections was established. Reputations grew.

The ground floors in the three twelve-story buildings were purchased jointly by all of the neighbourhoods to become retail and office space. Between two of these towers another building was erected to increase the concentration of retail at street level and to provide offices for workers above. The area between these buildings became a plaza. They had created a downtown.

One of these retail spaces became a café with comfortable seating spilling onto the plaza. The serious loitering that this encouraged added vibrancy to the core. Soon, some of the other spaces became a bakery, a grocery store, a barber shop, a hair dresser, a clothing store and a drug store. These businesses became the first choice for everyone because they were close by, owned and staffed by people from their town, and a place where they would be known, and not because they best answered a general desire for cheapness. Responding to this show of local support, other businesses that could survive on a client base of a few thousand people soon followed. Businesses that needed a larger customer base, such as a bookstore or a bicycle shop, were established in only some of the many towns that were now forming in the region. Most of what anyone needed could now be found in their town or an adjacent one nearby. Shopping took on a social dimension. These businesses created jobs that were filled preferentially by local townspeople, some of whom had previously been forced to commute away.

The town meeting that had organized the space between the neighbourhoods became the model for approaching other common needs. When better health care was needed, the town financed a clinic in the downtown core and hired a doctor. Concerns about security prompted the town to hire a bobby/mediator. Next came a few rooms to provide a safe place for young children wanting to expand their horizons beyond the neighbourhood. A town organization—a few permanent employees supported by volunteers—provided an ongoing structure. Everyone was engaged because people believed that they could actually affect the nature of this activity. They felt empowered. It became usual for civic ideas to be explored, differences to be uncovered, and compromises to emerge even before meetings took place. These discussions were generally informal, occurring over a cup of coffee or a glass of beer in the café, in chance meetings on a path, in the square or in an elevator. Projects were a response to real needs expressed by the community rather than being planks in party platforms. Community solutions could include both volunteers and changes in individual behaviours, options unavailable in distantly designed political solutions. The details regarding how individual lives can be changed by locally based health care, security and education will also be taken up in a later chapter.

More and more, the townspeople valued the unity and integrity of their community and recognized that the greatest threat to this was inequality. They knew they were stronger when everyone was involved and eager to contribute. They knew they were weakened when inequality bred discord. This gave urgency to their searches to find work for the unemployed, the underemployed and those with restricted abilities. When informal networks came up short, a town official was mandated to make sure that an appropriate job was found for each individual. The work week edged downwards as everyone was able to contribute. Identification with the town grew, inducing both volunteering and philanthropy. Jobs became willingly shared. Everyone's need for cash decreased.

Eve joined three other biking enthusiasts and opened a bicycle shop, building, selling, repairing and recycling bicycles—an increasingly common method of travel within the smaller ambit of the town. They took on the administration of a fleet of community bicycles dispersed throughout the town that could be picked up and left anywhere as needed. She loved her job. Control of her schedule, working less than three days a week, with a customer base of friends and neighbours, doing a needed task, she finally felt fulfilled. Her work had become an integral part of her life.

As lives centred on the town, cars continued to decrease in importance. Neighbourhood car sharing was extended to the town, where it was found that the one hundred and fifty cars necessary to provide a dozen in each neighbourhood could be replaced with thirty vehicles. Streets no longer had to be parking lots. With unobtrusive avenues allowing cars to access the busy streets on the edge of town, the interior streets could be further redesigned to serve people instead of cars. Much concrete and asphalt could be torn up, replaced with walkways, parks, trees, rose bushes and swing sets.

Up to this point, the towns had been growing unobtrusively within the city. True, it had been necessary to obtain building permits for renovations, but that was mainly a matter of ticking the right boxes with the help of the architect and structural engineer. Neighbourhood schools had been set up as private schools, but there are both precedents and channels for this. The clinic and bobby had to connect with established health and security regimes, but private clinics and private security arrangements also have precedents. These initiatives had all been carried out within the existing structure. They had been personal choices put into place using private resources. None of these activities were prohibited. But when the suggestion was to tear up city streets, they were definitely encroaching on someone's domain. Making a street disappear challenged the maintenance, planning and zoning authorities.

This was a political problem which required a political solution. If Eve's group were a unique enclave, then they would have to make do with what could be negotiated and turn their focus toward the next city election where they could advance the option of giving local political units wider autonomy. Changing the world requires that the idea of a cooperative life spread. That means trumpeting positive examples. In our story, however, we will assume that the change they had been experiencing was part of a general phenomenon. Other vibrant towns of families in neighbourhoods were coming together all through the region. The city administration reflected this evolution more with each election. City road crews showed up to help them refashion the streets. Other functions that they wished to address within the town, such as waste handling, also gained a sympathetic ear and a partner. New tools could be established while retaining the best of what already existed.

This area had been populated mainly with people who had assets and cars and apartments and jobs. Their affluence gave them the power to initiate change by choosing, buying, selling and funding projects. Deploying assets with purpose produces powerful systemic change from within. These possibilities existed because our society is wealthy and because the deal has spread that wealth widely. But not to everyone. There were people denied this form of agency either because of endemic poverty or tyranny. It was recognized by Eve, her neighbours and their new city allies that change needed to spread beyond affluent areas if they wanted to eliminate the seeds of conflict. They turned their efforts towards seeking out and eliminating barriers to equality.

Within their city, there was an area of persistent poverty. It had begun its life as cheap housing for exploited workers and this position had become self-sustaining as this substandard housing became the only possibility for the underemployed, the abandoned and the underpaid. This obvious need sitting on their doorstep was a challenge to their dream of an equal world. A grouping of the adjacent towns decided to make an equal place in the city available to these people.

The obvious need was income. There had to be enough well-paying jobs that one would be available to everyone. A manufacturer was persuaded to relocate a factory by providing a guarantee of support for three years. Two new city tools, one recreational and one professional, were also located in this area with the associated jobs going preferentially to the host community. Free, on-site training was made available wherever skills did not match the new requirements. Every new job had to pay a living wage. Since history had left a significant need for catch up, it was decided to augment each of these wages by half as much again, once more for a period of three years. This area of persistent poverty soon had full employment at a wage that left a surplus in the hands of everyone.

This surplus fed their desire for better housing. They formed into families, negotiated mortgages with their new allies as cosignatories, and began to plan. Because much of the housing stock had been badly maintained, rebuilding was often more effective than renovating. This activity generated more jobs which were also topped up. They seized the opportunity to share skills through apprenticeships. Money circulated, enriching everyone. Stores and cafés opened and thrived because there were customers with money to spend. This brought more jobs and more salaries which were also topped up. The tool that had been created by the rest of the city stood by to organize specific types of help when asked. When a way for locals to become doctors, teachers, engineers, electricians, carpenters and plumbers was identified as a need, training and mentoring was put in place. It turned out that putting money into the hands of poor people could eliminate persistent poverty. After three years, the situation was assessed. The top up to all salaries was replaced with specific programs targeted at a few lingering problems controlled by the townspeople themselves.

One interesting result was the transformation that occurred in a low-income project that had become rife with violence, crime, gangs, drugs and fear. This grouping of giant high-rises was still structurally sound so they did not have to be destroyed. Newly employed families flexed their new financial muscle, bought ten thousand square foot sections, and began to renovate. Groups of these families created neighbourhoods covering a few floors. These neighbourhoods made themselves secure by standing together to make it known that troublesome strangers were unwelcome. When fifteen of these neighbourhoods created a town out of one of the huge towers, they were able to hire a bobby to extend the zone of safety into public areas. Since everyone could now see a path to a well-paying job, gang membership plummeted. Aimlessness and violence was replaced by neighbourhood vibrancy. Work bees turned the space between the buildings into a park. Affluent customers allowed stores to thrive in the ground floor of one of the towers. No outside agencies had planned and directed this change. People were able to transform their own lives when they had paying jobs.

Groups of towns got together to produce tools. Some were the old city agencies which were restructured to assure control by individuals. Others were entirely new entities. Some used fee-for-service; some needed other constructions to keep them locally controllable. Some generated conflict with distant levels of government and the implementation of local control had to be negotiated.

Ideas spread quickly in the digitally connected world. During the first few years that Eve's group was renovating, establishing towns and local businesses, and building tools, they were in contact with friends and friends of friends all over the city, in other cities in the region, in adjacent suburbs and rural areas, and even on other continents. Other groups were inspired to follow their lead and engage in the same exercise. Every success spread the idea and seeded new groups of families. There was no rollout, no plan, no organization. Information spread, examples were noted and simultaneous activity changed the world. Links between towns, regions and continents were only established after the change had occurred.

The most significant and intractable barrier to eliminating inequalities was with people who were not free to make their own choices. Our feudal past has left many at the mercy of kleptocrats manipulating corrupt governments. Transfers to people in autocratic regimes will surely be stolen. Trustworthy information cannot exist because the handmaiden of domination is lies. The world hardened into two blocks—one resting on empowered individuals made up of North America, most of South America, Europe, India, much of southeast Asia, and a few countries in Africa. Meanwhile, in Russia, China, large parts of Africa, much of the Middle East, and scattered regimes in southeast Asia and South America, tyrants were able to keep people in thrall.

It was impossible to responsibly deal with these tyrannies. Their products in trade hid various types of exploitation. Anything transferred to them would be used to fortify the elite against their own people—withholding necessities is a powerful weapon. Kleptocratic money was always tainted with suffering.

As transfers eliminated poverty throughout free regions in the developing world, digital connections made this known to people in the tyrannies. They knew that there was an alternative to their constrained lives. Tyrants had increasing difficulty recruiting and motivating soldiers. People became harder to bully and more likely to object. In each state, the moment grew closer when the soldiers went home to their mothers and the kleptocratic apparatus collapsed. The energy this sudden flowering of freedom unleashed, backed with the readiness of transfers from the free communities, allowed these states to bloom once the parasites were removed. Resources were redistributed, local businesses thrived without the kickbacks, everyone gained useful work, long term plans could be made without the fear that all gains would be stolen. One by one, lands with people keeping their heads down were replaced with vibrant, free communities.

A global body came into being to address problems common to all. As equality spread, there was no longer any impetus to war. Nuclear weapons were gathered up, turned into electrical power where possible, and otherwise kept carefully isolated. With global cooperation, hunger was eliminated and endemic diseases were controlled.

Ten years after their first meeting, Eve and her family gathered for their regular anniversary dinner. The group had changed. There had been a death—unexpectedly one of the youngish singles had developed a fatal brain tumour. Three people had left the family, finding other partners and other families in other regions that better fit their unfolding lives. Two new people had entered their group from outside and there had been two births. Their wage work commitment was low and still dropping. Their lives centred around home with time for art, writing, travel, music, learning and volunteer civic efforts. Exuberance surrounded them.

The world looked different on their news feeds. Hunger no longer existed. War was an abomination relegated to the history books. The Doomsday Clock, instituted to record the imminence of nuclear annihilation, had been dismantled. Simple lives and responsible energy production in every neighbourhood had reduced carbon emissions to a fraction of their former selves and that trend was picking up speed. As food production became local and as food choices moved away from beef, green spaces increased in many parts of the world, both sequestering carbon and restoring habitats for many struggling species. Eve smiled as she watched her daughter sharing a book with one of the seniors, both giggling away. She felt proud of the world she could bequeath to her grandchildren's grandchildren. She only faintly remembered what it felt like to be depressed.

CHAPTER TWELVE

Work

∿➤

Wage work will always be a part of our lives. There will always be things that need to be done and we will need a way to self-organize the allocation of this work. The most efficient way to do this without an overseer is to have people apply for jobs and get hired to do that task. And we will all need a wage to allow us to purchase those things we want and need. Responsible consuming is the motive force for self-organizing the production and distribution of goods and services—this requires a wage in every wallet. But though jobs will continue to exist in the new world of families, neighbourhoods and towns, they will look very different than they do today.

When we work where we live and shop and play, our lives will regain the unity that existed in the pre-automobile world. Those staffing our cafés and stores, the teachers in our schools, the bobby walking our streets, the carpenters repairing our buildings, the barber cutting our hair, the electrician or the plumber making house calls to resolve our crises, the doctor and nurses in the clinic, the mailman, the shoemaker, the landscaper, and all others serving our needs, will be known residents of our town. They will be the same people we meet casually in the café, encounter in the gardening club or join with to craft solutions in the town forums. Every interaction will strengthen the web of connections that surrounds us. As we regularly deal with people we know, friends or friends of friends, known personally or by reputation, a community is formed and strengthened.

Jobs can be done more effectively when workers know their clients. For example, a teacher with a pupil who is acting out can respond more appropriately if he knows details of the child's life that don't appear on an entry questionnaire. The experience for both clients and workers is richer when commercial interactions have a social dimension. The café and the barber shop become more than places to refuel or get your hair cut as people meet, gossip, joke and discuss. The school is less daunting to young children when they already know the adult who is to be their teacher. Healing is more effective when doctors and nurses are also friends. House calls by the plumber or electrician build connections within the town as well as within the walls. When neighbours rely on our work, we take more pride in the doing.

If a particular skill, product or service is lacking in our town so that we are forced to purchase it from outside, then we should see if we can get that function filled within the community. For example, recognizing a need for a qualified teacher, policeman, or doctor, we could sponsor one of our own to take the appropriate training. In the interim, we could hire from outside on a temporary basis or hire from outside with relocation being a condition of employment. We want our money to stay in the community. Local people filling local positions should be an important determinant in our purchasing decisions.

Many jobs in large tools can be filled remotely by workers dispersed throughout the market area. If we are funding a tool, we should expect that some of the jobs so created should be located in our town. A major concern in this arrangement is the isolation that can result from the demise of the work group as a social entity. However, if a group of people, each working for a different tool, share a work space in the neighbourhood or the town, then employees would have supportive coworkers who were also friends in other facets of their lives. These are connections that will continue regardless of retirement or job changes.

More local people who are working remotely while based in the community would add life to our streets. While workers would be connected to their tool technically, much of their support—reception, delivery, copy services, stationary supplies, IT support, lunch—could be sourced in the town, permitting a wider range of business to thrive locally. This would benefit us all.

An argument against remote employment from the point of view of the employer is that it eliminates the serendipitous discoveries that can issue from informal interactions. This is so valued by some businesses that they have designed their workspaces to encourage these interchanges. They are lost when employees connect only through email and Zoom. This may be significant for a few tools but a compensating gain will arise in most neighbourhood workplaces. If a remote worker shares their workspace with ten neighbours there will be a more varied group in the room which is potentially more fecund. Instead of technical people meeting informally with other technical people, a technical worker may now be sharing office space with a designer, a graphic artist, a marketer, a writer or a technical person from a different field. This variety of perspectives is also likely to produce those unexpected eureka moments.

We all gain in another significant way with dispersed institutions. As tools become large, transparency is more challenging to engineer as clustered employees tend to protect their power by keeping secrets. But when employees are dispersed, each is a member of a community as well as a member of an organization. They become a voice who can explain their tool to the community. They become eyes and ears providing feedback to the tool. The dispersed tool will thus have a real presence in many communities. It will be closer to its users. This will benefit both the tool and the community.

Dispersed institutions will use facilities that already exist to meet other needs. Instead of having to produce and staff a cafeteria, workers can use the village café, where their patronage will spur more offerings, more profits and more jobs. Instead of using time, roads, cars and parking facilities commuting, people can walk to work, spending more time in their homes and on the streets of the neighbourhood. Those working from six to two or from seven to three will be around in the afternoon. Those working ten to six will be there in the mornings. There will be people present if needed to respond to emergencies, for casual conversation or for pickup soccer games.

As we do our job, we become necessary to each other. These activities bring a community together. If some people do not work, it draws a line between workers and consumers. Young children should be exempted for reasons of competence and safety. Those who are ill or for whom movement has become painful should be excused. However, youth, elders and the disabled should, as much as possible, be helped to find positions commensurate with their abilities, strength and energy. As soon as children can perform useful work, they should be encouraged to do so and provided with real wages for their efforts. This money will permit them to contribute their share to their family's food and rent as well as having a say in the marketplace as discretionary consumers. As they grow older, they will progress to more strenuous and complex jobs until, in their teenage years, they are contributing fully for a full wage. Elders could progress from more to less physically demanding roles, allowing them to stay connected, contributing and financially equal while allowing the community to benefit from their experience.

If everyone controls a living wage, then everyone can reflect their beliefs into the world. Our world should display both the wisdom of elders and the exuberance of youth instead of both being relegated to some antechamber of life. For youth, having the responsibilities of earning, of paying their own expenses, of being taken seriously in discussions because of their financial stake, will be a maturing experience. For elders, the continued engagement will maintain their sense of purpose and give the world the benefit of their experience. When some perform the work while others live on those efforts, equality is undermined.

When someone needs a job, either because they are looking for a position that is more challenging, because their job is not correct for them, or because their job has disappeared leaving them unemployed, then it is important to the preservation of equality within the family, within the neighbourhood and within the town that the search end in success. When one person is unemployed while the rest work, the community is weakened. That makes unemployment a community problem. This is not a new realization. Family members and friends today help family members and friends get appropriate jobs because everyone knows how corrosive an inability to contribute can be. When someone needs a job and her own efforts have not borne fruit, other members of the family will work their contacts. That may very well solve the problem.

But if the family cannot produce a job, then the neighbours must be engaged. They are familiar with the job seeker's skills, abilities, personality and reputation so their efforts will begin at a point well beyond a resumé. Amongst these one to two hundred people, there will be many connections to employers, clients, suppliers and tools. Introductions can be made. Neighbours can put themselves on the line vouching personally for the aspirant.

If this search still comes up blank, then the underlying situation needs to be examined. Perhaps the job seeker needs more training. In this case, the neighbourhood can offer encouragement, tutoring, mentoring and financial support. Perhaps the individual has work habits or a psychological stance such that no neighbour would vouch for him. This barrier to equality must also be faced and the family and the neighbourhood are the proper places to do so. If it turns out that there actually is no slot that fits this odd-shaped peg, then it may be necessary to create a job specifically tailored to the individual. If they are valued as a member of the neighbourhood, then all avenues must be explored.

If we still come up short, then we must take our search to our contacts in the town. For example, if there is an interest in the healing professions, then we can approach the local doctor for information about jobs in the medical sector. Similarly, if the interest is auto mechanics, bookselling, food growing or whatever, we approach our local source and, through them, look for a position, an apprenticeship, or, at the very least, accurate information and further contacts. It may also become evident that there is a need for a more formal service in the town administration to help match job seekers with appropriate jobs whenever the informal network fails. Finding the right job for everyone is an important community task.

If, however, no jobs can be found because there is an absolute shortage of wage work, then it will be impossible for families, neighbours or the town to be a bridge to useful work. Our quest for equality demands that, in that case, those who are already employed should share their jobs. To frustrate equality by holding onto a forty-hour work week when others have no job at all is a form of hoarding. Sharing jobs may require retraining, and it can create scheduling challenges, but training and scheduling are problems that must be dealt with in any system of employment.

When we all know that jobs will be shared, they cease to be prizes in a competitive struggle. Our continual striving against our fellows for better jobs, higher wages and more hours is one of the main drivers of growth in the machine. Approaching our work with the attitude that we are all sharing the tasks that need to be done will allow our economy to assume an entirely new shape. Instead of job-seeking stimulating competition and creating growth, as is the situation at present, we would be able to generate a steady state or even a shrinking economy if either were required. Growth would be the choice only when it is demanded by our needs and demographics.

The threat we have lived with for two centuries that a shortage of new markets will cause glut, failures, lay-offs, and recession will disappear as output can be painlessly reduced by having employees leave or cut their hours in the secure knowledge that the community will produce alternate employment. The lurking danger that artificial intelligence and robotics will make most workers redundant, intensifying battles for a shrinking number of jobs, will be changed from that negative to the positive promise that these new technologies will liberate people from some of the wage work required to make society work. The workplace is where our ability to cooperate truly becomes transformative. This is where "the threat" is defanged. And the bonus is that everyone's work week will continue to decrease as we share the work.

Organizing this sharing of work over a large population can be easily arranged. Once again, the crux is good information transparently disseminated so that we each can know what constitutes a fair share of the necessary work. That calculation requires only two things—a good estimation of the total hours of work required to fill all the jobs in a society along with a count of the number of people who will be contributing. Neither of these are as complicated as the many measures of the economy that are now gathered in a vain attempt to predict the future of a machine that is out of control. An independent, trusted agency should be able to produce a number which describes the average wage work commitment expected of each individual and to monitor how it changes.

With this number, we can each adjust our own behaviour. If we are working above this number, a desire to be responsible will induce us to ask for one less shift; if our contribution is below, we can ask for extra work. The activity of anyone working hours significantly above or below this number will be obvious to the community—they will be seen either as Hoarders or as someone who could use help in finding more work.

Though the hours we work will decrease, everyone must still take home a living wage. This implies that every hourly wage will have to increase as the hours of work go down. This should not be a problem. Consider the following example. If 30 percent of the costs of a business previously went to wages and each person was replaced by a group of three, then labour costs would triple but total costs would only rise from 70 plus 30 to 70 plus 3 times 30. Prices could be expected to rise about 60 percent. Labour costs are not the only determinant in the price of a good or service.

However, wage costs would not actually need to triple to maintain a living wage because there would be a corresponding change in what is meant by a living wage. In each family, there will be more earners than before as the youth, the elders and the previously underemployed are able to contribute. More contributors would mean that the portion of expenses that must be carried by each individual will drop. Having time to do things for ourselves that we previously had to purchase will further reduce the amount we need to earn. In our example, each of the three salaries could be considerably less than the hourly wage for one person doing all the work. Expected price rises need be considerably less than 60 percent.

Also, some costs will go down in this new arrangement. Large differentials in wages earned will be more obvious as we take action to share our hours of work. Having some people earning a minimum wage while others are claiming a pay packet ten times that amount is another negation of our striving for equality, as serious an act of hoarding as claiming hours while others have none. If this persists, then an equal future based on cooperation will be impossible. It must be challenged. There are some justifications for salary differentials for dangerous or unpalatable work, or as compensation for hardship or arduous training periods, but the differences cannot be huge without risking the creation of classes of haves and have-nots. Our present gulf between the well-paid and the poorly-paid is a carryover from a time of class differences which has persisted mainly because of systemic feudal advantages such as preferred access to education and networking. As everyone chooses to share, and as we move towards equality, large wage packets will decrease. This will cause some costs go down, further reducing what must be considered a living wage.

There may be a need for community mandates during this transformation of the workplace. For example, it may be necessary to institute a maximum wage as well as a minimum wage. Also, if people are reluctant to give up hours, it may be necessary to introduce a graduated pay scale—full wage for hours up to the fair share, half wage for the next period, then quarter wage for hours beyond that and so on, associating a cost with the hoarding of this scarce resource, incentivizing a reduction of hours worked without denying the choice to work. It would be hoped that changes would all flow from responsible choices and a desire to increase our leisure time, but change can be difficult, and periods of adjustment may require a little push.

As our wage work commitment decreases, we will be able to give more of our time within the family and the neighbourhood. Preparation of food for others is a gift—we will gain the time to do this without being rushed. Doing the dishes or washing the kitchen floor are also gifts. When we eat in restaurants, these functions are shifted from the gift to the cash economy. Caring for children in a family and giving support to a primary caregiver are gifts: purchasing day care moves these acts into the cash economy. Caring for elders as they decline is a gift. That too has lately been transferred into the cash economy with a notable lack of success. Moving items from the cash back to the gift economy reduces both the total wage work a society must accomplish as well as the size of the wage packet required for a good life.

Currently, families are so small and hurried that the space for these gifts disappears. When we compare the options of a single parent with one child to those of a family of two parents with two children, we see a great increase in capacity that arises from the ability to share chores. And when families are further extended with grandparents, adult siblings and close friends nearby, the myriad interactions and freely given time again increases the ease with which all can meet their needs. In an empowered extended family of around ten varied individuals, the gifting will become quite powerful. We will be able to meet many needs within the family for which we now are forced to turn to the cash economy.

We also will be able to be generous with our time in the town, volunteering in various bodies and giving time to help provide things that had previously only existed in the wage world. A town with shared goals and activities and an active meeting space that encourages working together will be enriched by these gifts.

So exactly how much wage work and gift work will we have to contribute to produce a good life? Firstly, as we do more things for ourselves, for the family and for the neighbourhood that we previously had paid someone else to do, we reduce the total amount of wage work that needs to be done. Consider our kitchens. If we were to make some of our own bread, granola, jam, mayonnaise, cookies, cakes and meals instead of buying them, we remove demand from the factories that produce these items. As we contribute to child rearing, eldercare, gardening, cooking, cleaning, building, fire safety, security, health care and other facets of our lives, we invigorate our families with activity while reducing the need to pay other workers a wage to accomplish the same things. Doing for ourselves will reduce the total amount of wage work.

Secondly, the market for many things we produce today will decrease or disappear. The slowing to a stop of population growth will reduce demand for cars, furniture, houses and much more, freeing up much labour.

Buying as a balm to soothe our loneliness will decrease as we become embedded in nurturing families. Since it will now be obvious that lower levels of consumption leads to a shorter work week for all, the demand to have more things will be countered by the desire to have more time. Needing less and doing more for ourselves will cut the wage work required by as much as half from the situation that exists today.

Thirdly, each person will need to do less wage work when everybody contributes. Currently, the unemployed, the underemployed, the able retired, those disabled in some capacity but still able in others, the youth, and those who have just dropped out of the counted categories, are denied a chance to contribute. In most of the developed world, this non-working segment adds up to more than a third of the population. In the developing world, it is even greater.

Fourthly, as entrenched agencies are transformed into tools which must prove their utility to earn our continued support, their high labour requirements will decrease. We will choose to dissolve some of them completely in order to approach problems locally, drawing on local resources and the efforts of local people who have a stake in the game. The apparatus for collecting and redistributing taxes will be largely unnecessary when decisions on transfers and the funding of tools are returned to the individual and the town. Then there are those parts of government that have served their purpose but continue to live as zombie corporations because there is no mechanism to make them disappear when the need for their service has passed. It is possible for large groups of people to look busy, pushing paper, spending their energy on holding meetings, regularly reorganizing their structure and issuing reports that are carefully filed, while producing no actual useful output. These workers will be pleased to trade this life for a much shorter week spent doing the useful work they will be promised in this new regime.

When people are asked on surveys if they think their job is actually useful, the responses are illuminating, if depressing. A 2018 interview in *The Economist* with the late anthropologist David Graeber from the London School of Economics, who authored the book *Bullshit Jobs*, began with the following exchange:

> *The Economist:* What is a bullshit job?
> David Graeber: A bullshit job is one that even the person doing it secretly believes need not, or should not, exist. That if the job, or even the whole industry, were to vanish, either it would make no difference to anyone, or the world might even be a slightly better place. Something like 37-40% of workers, according to surveys, say their jobs make no difference. ... we should proceed on the assumptions that for the most part, people's self-assessments are correct. Their jobs really are as pointless as they think they are.[1]

These jobs persist in both government and industry because people must hold onto the wage, not because the output of their efforts are necessary or useful. However, when everyone knows that no one will be left without a job when these positions are identified and eliminated, then people will cooperate in removing these inefficiencies. Finding ways to quit wasting effort will replace the impulse to create a smokescreen to disguise uselessness. This should reduce at least another third of the wage work we presently do.

When enough consumers pass on a product, then production facilities must be scaled back. In a society that shares wage work, such an event would be applauded.

Instead of a closure meaning unemployment and distress for a few, it would mean that these workers would be helped into new positions while everyone's expectation of wage work would decrease. Everyone benefits so everyone will help wring waste out of the economy.

When wage work is shared by the whole population, the forty-hour work week will drop by a quarter to a third. When we eliminate useless goods and bullshit jobs, it will drop by another third. When we move our efforts from the wage economy to the gift economy, work commitments will drop drastically again. When innovation becomes focussed on liberating hours of work by more efficiently producing those goods we do need, there will be another gain. It is hard to imagine that each person would have to work more than a couple of days a week. This commitment should continue to decrease. One day a week spent in wage work should not be out of sight for each of us.

Depending on the requirements of the job and the desires of the individuals involved, wage work could be organized as one long day a week or a couple of short days, perhaps one week on followed by two weeks off, or three months on and six months off or whatever arrangement would allow someone to contribute a fair share while still making choices to create their best life.

This would allow two, three, four or more individuals to share what constitutes a job today. When they all live in the town, with their workplace also in the town, they have incredible flexibility. They could share their schedules to accommodate their other enthusiasms or family demands. Such a team could also match their schedules to the demand for their service. It is impossible to ask one employee to appear twice on Tuesday and not at all on Friday, or to work twice as long through September while reducing their hours through the slow period of March, whereas that type of scheduling would be easy for a team. Night shifts would become a less damaging feature of life if they only occurred a couple of nights every third or fourth week. These teams could incorporate the wisdom of age and the energy of youth as they shared their tasks as a team.

Eve and her partners in the bicycle shop were very pleased to expand their numbers from three to eight and to reduce their work week from four days to a leisurely two. She appreciated the chance to spend more time with one of the older family members whose health was declining—the family had organized that this person was never left alone. She also signed up for the music lessons she had never seemed to have had time for and, as her skill progressed, found a group to play with one night a week. Her life was changing, becoming more satisfying.

Presently, the only way we can gain security against poverty, losing our health, or losing our job is by accumulating money. Such fears of misfortune are a large part of what drives us to work long hours and to hoard our wealth. But this competition to pile up our own security generates both inequality and the problems of growth and this, in turn, stokes the envy, fear and violence which played a large part in making us feel insecure in the first place. It is a negative feedback loop built on a false premise. Money as a hedge against disaster has proven to be extremely unreliable as many who have suddenly found themselves facing depressions, inflationary periods, wars or natural disasters have found. True security can only be created by embedding ourselves in a community committed to sharing. As we come to value the security that flows from equality more than the security that comes from money, we will be increasingly eager to share our work. In doing so, we gain time while becoming more secure. We stand to lose nothing except our long hours of work.

A person will isolate himself from this vibrant community if his wealth only benefits himself. Displays of wealth will be a barrier to inclusion. As reputation becomes the currency to be valued, philanthropy will gain in importance. Enhancing the reputation of family, neighbourhood and town will become an important use for accumulated wealth.

Being included is the only true security. If we become unemployed, our community will help us find a new position. If shortages occur, then we know our mates will share what they have. Should we become ill, we can count on care within the family and the neighbourhood. As we age, there will still be a valued place for us.

A willingness to share our jobs will transform our strivings from fueling growth into the nurturing of community. But striving will continue. It is the nature of humans to be curious, to be the Originator, to seek the esteem of our peers. However, when we deflect that striving away from wealth accumulation, we defuse the threat of unstoppable growth.

Our obligations to gifting in the family, neighbourhood and town must be taken seriously. These could be expected to demand another couple of days a week of our time. Fourteen people each offering two days a week adds up to a very significant contribution to family and community health. Chores in the home such as cooking, cleaning, laundry, childcare, elder care, repairs or gardening could be done leisurely and purposefully instead of being taken up when we arrive home tired, or need to be crammed resentfully into a busy weekend. Homes and neighbourhoods would hum with the activity that turns a house into a home. Volunteering in town activities such as health care, fire protection or other town organizations would help us produce solutions to shared problems while we invigorate our community.

Our wage work and our gift work should easily be done in less than four days a week. That estimate is likely to continue to decrease. Beyond this, our time will be completely our own. We will be free to follow our interests in conversation, travel, sports, games, drama, study, music, art, crafts, decoration, writing or invention. In the past, whenever the Originator was given time in an atmosphere of freedom and equality, there followed an explosion of creativity. There is no reason to believe that recreating this setting would have different results today. A new golden age will flow from our commitment to sharing our jobs.

Chapter Thirteen

Responsible Infrastructure

⌁➤

Sharing our jobs replaces the Dominator with the Cooperator at the heart of our world. When we do this, we are embracing a new way of approaching the world. New tasks will fill our days. New pathways to esteem will arise. New satisfactions will drive our activities. New taboos and traditions will evolve to add structure to our daily lives. Committing to cooperate allows us to grow into a whole new survival strategy.

Like any survival strategy that hopes to persist long-term, it must guarantee food security. It must allow us to meet our needs for energy without causing harm. And it must allow us to deal with our waste in ways that are not damaging to the planet. Our present strategy fails to meet these tests. These failures are incubating serious long-term dangers. In this chapter, I will look at how our new arrangement will allow us to meet these needs sustainably and responsibly. Because the urban, suburban and rural situations present quite different challenges and opportunities when it comes to infrastructure, we will consider them separately.

The Urban World

The world has become increasingly urban over the last century. From a situation a few hundred years ago where 90 percent of humanity lived on the land and tilled the soil mainly with human power aided, if they were lucky, by a draft animal, now over half of humanity lives in large cities while the land is tilled by a few percent of the population with the force of their labour magnified many times with machinery. The flow of people from the fields to the cities seems unstoppable. The exuberance and variety created by density, the increased interactions, and the greater opportunities for employment are powerful draws for people who have been living lives of isolation and poverty. The possibility of remote work may reinvigorate rural areas, but it seems likely that urban spaces will continue to be home to most people in the future.

Urban housing stock is currently a mixture of high-rises, low-rises and areas of closely spaced, single family homes, perhaps detached, perhaps not. A space adequate to house a family of ten to fifteen—around ten thousand square feet—could be a floor or two in a high- or low-rise, or a few detached homes with connecting and infill structures on a residential street. A neighbourhood could be part of a high-rise, a few low-rises, a residential block or mixtures of these. Towns would likely include all three types of housing stock though any one may predominate.

Life today relies totally on the availability of cheap energy. The invention of methods of transforming falling water, wood, coal, fissionable materials, and oil and gas into electricity which could be delivered to us over large distances spurred the construction of the modern world. This availability of electricity is what lets us live in comfort. Unfortunately, this bounty has had a cost: though it may be cheap for us to purchase, there is a large deferred cost accumulating in damage to the commons. Flooding the valleys that sit behind our hydroelectric dams, mining and burning the coal, levelling our forests for charcoal, creating ever growing piles of nuclear waste, producing carbon dioxide by burning fossil fuels—all have left the world more tattered and vulnerable. But we are addicted to this energy. We demand that cheap electricity production continues to multiply to serve our growing appetites. How can we attain the energy we rely on to build, heat and light our homes, to power our transportation, to run our stoves and ovens, hot water heaters, coffee makers, refrigerators, computers, televisions, phones and a million other items, in ways that do not damage our world? Personal responsibility demands that we stand behind every action we take, every light switch we throw and every mile we drive. We must gain control over the tools that provide for this part of our lives.

There are two ways to approach this question: we can make our power production cleaner so we can continue to consume at the same rate while reducing the harm or we can reduce the amount of power that we use. We need to do both. Empowered family units will inherently need less power as ten to fifteen people share heat, light, ovens—everything. Also, in the renovation of our space, we could make energy efficiency a priority: insulation of walls and windows, integration of thermal mass to store solar heat, and the installation of heat exchangers are more reasonable investments for groups of fifteen than for groups of one or two. Every demand we make on the power grid must be examined for ways we can reduce it if we wish to take responsibility for the hidden costs. There have been great advances in materials and design for energy neutral buildings and energy efficient appliances over the last few decades. We should incorporate these features whenever possible. There are sizable gains we can seize. Taking personal responsibility for the deferred costs has been the missing ingredient.

Increasing the amount of work done remotely reduces the need for both automobile production and automobile miles. Fewer cars need less steel. Fewer miles need less fuel. Living a simpler life means that we consume less of everything, and this in turn reduces our demand for power. Shaping our activities in the neighbourhood and the town as responsible consumers will allow us to demand minimal energy use in every proffered solution. Keeping the deferred costs of energy in mind will have an effect.

But reducing usage can be only a partial solution. Our lives are built on cheap available power. Returning to a reliance on human muscles and draft animals would end the ease, leisure, comfort and possibilities that we have achieved. Our ability to support billions on the earth would disappear, leading to starvation. We need clean power.

When we rely on distant, massive, secretive, uncontrollable sources for our power, then the harms become problems that are separated from our act of demanding the power. Cleaning up the pollution that is due to our use of energy becomes someone else's responsibility. We can mentally disconnect the damage from our decisions. But make no mistake: we are the responsible agents. The problem of powering our lives is our own.

As the problems of energy pollution have become more dire, investment has flowed towards innovations to generate clean energy. Our options are becoming more varied and more affordable by the day. Much of this investment, however, is coming from the large energy companies who fear a loss in their current market share when we reject dirty fuels. They are searching for large-scale clean energy solutions that they can continue to monopolize. It is crucial that we generate the demand that will encourage small-scale solutions that can be implemented at the family, neighbourhood and town levels. When empowered individuals, families, neighbourhoods and towns are the decision makers and the purchasers, innovation will be directed towards small-scale, responsible, sustainable solutions. New options, currently unavailable, will start to appear.

Solar is a dispersed, renewable energy source where the costs occur in the mining of the necessary rare metals and in the manufacturing process itself. These costs can be reduced by responsible consuming—obtaining verified information and making responsible choices. But the sun shines on every square inch of the earth for a part of most days and every building has a roof. We can all make use of it to capture some power. In sunny regions for groups with a large roof area, this, coupled with battery storage, may be enough to meet their needs. However, in urban areas with tall, multi-unit buildings, the rooftop to resident ratio is so reduced that solar is probably going to be no more than a partial solution.

If there is unused land nearby, the town could explore the possibility of establishing a solar farm. As it will be close, the disruptions of transmission lines are reduced and, servicing only a town, it could be kept small enough so as to not overwhelm an area. The consequences of our demands would be visible to everyone during their regular ramblings. Harnessing wind or falling water may be possible in rare situations, but they would be impossible in most towns.

The only other technology which can create power without ongoing emissions within the boundaries of a town is nuclear. If locally sited solar cannot be made to meet our needs, we may have to look at this option. The history of nuclear energy is inextricably bound up with the creation of the atomic bombs that were detonated over Hiroshima and Nagasaki at the end of the Second World War. Subsequent nuclear development during the Cold War was skewed away from safe power production towards reactors that could produce weapons-grade byproducts for the military. Innovators were never able to focus on safe power generation, let alone small-scale, safe power generation.

Meanwhile, disgust at nuclear weapon proliferation shut down most nuclear research and development. The advances that would have made this type of power generation safer, cheaper and more efficient were never allowed to occur. Many people rejected everything with nuclear in the name with the unfortunate consequence that, as our need for power grew, that need was met almost exclusively with coal and oil. Our success in shutting down nuclear power sped up the accumulation of greenhouse gases, perhaps something that will come to be seen in the history books as a pyrrhic victory. However, one development program for small-scale reactors did continue, funded largely by the need of the military for submarine, aircraft carrier, and other remote power sources that would be able to go long periods without refuelling. As a result of this work, there are now designs for a type of reactor that could safely produce power without emissions for an entity the size of a town.

These are known as small modular reactors. Small units, capable of powering anything from a town to a city, could be produced on an assembly line and trucked to their eventual location where they could safely produce power and hot water for decades with minimal servicing. Users would be able to receive power with no emissions with both use and generation contained within the range of "themselves or their doings." The only remaining problem would be dealing with exhausted fuel at the end of the cycle.

These reactors are safe. The designs all incorporate automatic passive shutdowns. Being smaller, they can more easily be enclosed in large containment structures, possibly underground, if further isolation is felt to be necessary. They are a different animal than large-scale nuclear which, in itself, is still much safer than any other form of energy production. When we consider mine disasters, oil rig collapses and spills, gas leak explosions and the health consequences of polluted air, it is clear that no energy source has ever been completely safe. Death rates per terawatt hour from accidents and air pollution show nuclear causing 0.03 deaths, natural gas forty times higher at 2.8, oil about six times higher again at 18.4, and coal topping the list at 24.6.[1] These figures include the Chernobyl disaster, something caused completely by political incompetence. The reflexive desire to reject everything with nuclear in the title must be looked at in terms of our urgent need to stop harming the planet. It may be the only responsible local choice for a town.

Another design which could help solve the problem of nuclear waste—the travelling wave reactor—is also once again active in the development stage. These reactors are large and capable of powering a large area, so they would have to be organized as a tool. They use a small core of enriched fuel surrounded by a much larger pile of lower grade, fissionable products. Radiation initiating in the core gradually converts the lower grade products into fuel, which then creates the heat to drive a generator. The reaction gradually travels through the low-grade fuel, continuing for decades and stopping only when the fuel is exhausted. A benefit beyond the gaining of power is that the low-grade portion of the fuel could be made up of spent fuel from power plants while the high-grade core could come from decommissioned weapons, both now considered problematic nuclear waste. This reactor turns nuclear waste into power, leaving a much smaller and less potent residue at the end of its decades of use. According to Wikipedia,

> Terrapower (a company currently developing a design for travelling wave reactors) has also estimated that wide deployment of TWRs could enable projected global stockpiles of depleted uranium to sustain 80% of the world's population at U.S. per capita energy usages for over a millennium.[2]

Our stored nuclear waste is a huge storehouse of emission-free energy. Though no such plant has yet been built, the design is feasible. The major barrier is that governments are terrified of the political risks inherent in the word nuclear so they have been refusing any permissions to build a working model.

If we did decide to establish a small modular reactor to provide power for our town, it should not be hidden away. The town organization responsible for its safety must have continual access and the power to make decisions. If the citizens around Fukushima had had unrestricted access to the power station, surely someone would have noticed that locating the emergency power generator in the basement of a facility in a tsunami zone was a bad idea. If citizens had been on site at Chernobyl, the risky test with untrained personnel that precipitated the disaster would not have occurred nor would critical hours have been wasted before calling in help as those responsible watched the unthinkable develop as they dithered about the political effects on their careers and their lives. Hiding a facility behind a fence and staffing it with salarymen virtually guarantees irresponsibility.

A variety of designs for emission-free power currently exist, in a variety of sizes and outputs. More options will soon arrive. We can meet our need for energy within the world of "ourselves or our doings."

There will be many fewer cars. Car sharing would be the responsible transportation solution where fifty to one hundred vehicles could easily service a town of a few thousand. This reduction in vehicles by an order of magnitude would save both steel and fuel. With electric cars, the driving we do can be as clean as our energy source.

We would also only need parking for this number of cars. Ten to twenty parking spaces in each neighbourhood should be more than sufficient. These already exist either beside or under our high rises and low rises. For row houses or detached homes, there would have to be some small lot at the edge of the neighbourhood. We can make cars an unobtrusive part of our lives rather than their most significant feature.

Minimizing the use of cars like this would revolutionize our streets. There would no longer be rows of parked vehicles separating each building from the next, making noise, fumes and accidents an inescapable part of our lives. There would be access provided for emergency vehicles and deliveries but, with parking deliberately sited to provide quick access to the busy streets that connect towns, there would be few cars within a town. Streets can be turned into landscaping, paths, benches and play areas. We can live in a park.

As the number of cars in the world plummets, the automobile sector of the economy will claim fewer resources and fewer jobs for manufacturing, repairs, servicing, road creation and maintenance, and gasoline extraction, refining and delivery. Every part of this huge industry uses energy and the costs are measured in dollars, hours of work demanded and emissions created. The savings in all these areas will be considerable, showing in the money that is left in our pockets, in the shortening of our wage work week, and in the carbon dioxide equivalents that are not spewed into the atmosphere.

Vibrant towns joined by roads that connect them without going through them define a grid for a transit system. Currently, much public transit is designed to offer low cost, low efficiency, low comfort, low enjoyment mobility to people who do not have a car. These criteria have led us to large vehicles of minimal comfort on infrequent schedules, often either jam-packed or virtually empty. However, if transit were to be designed to be attractive and useful to everyone, the client experience will become central. We will end up with smaller vehicles on more flexible schedules that can be made to mesh more seamlessly with our lives, with comfortable seating placed to make human interactions natural and enjoyable. Transit will come to be an extension of the public realm of parks, benches and walkways. Some routes would connect townspeople directly to specific locations. For example, there may be a need for regular access to the city centre, to a nearby metro station or to the university. The metro would be a tool designed to provide rapid access across a large distance, but the client experience would still have to be paramount. Transit designed to attract people could make it more pleasant than driving. Success here would further reduce the number of cars which must be kept in each neighbourhood, further saving resources and further liberating our streets.

Responsibility also demands that we deal with our own waste instead of looking for an out-of-sight, out-of-mind solution. We currently flush our toilet and watch our waste swirl away to become someone else's problem. We fill our bins, put them by the curb and assume any problems associated with that waste are solved. The real effects of sewage systems such as wastage of water, contaminating leakage from sewage pipes, offensive odours and environmental disruption where it eventually must be treated, the disposal of any by-products, and the chemical additives necessary to the process remaining in the environment are still our responsibility even after they have been moved out of our control. We must be able to see the damages created in our name so we can find ways to deal with them.

There is so much installed infrastructure that sanitation professionals must continue to patch rather than look at alternatives that may question the wisdom of their pipes, pumps, processing and expertise. But, here too, there are feasible alternatives that can be local. We can deal with our own waste in ways that are effective and in ways that can guarantee we will affect no one else's life. It turns out they are also safer and use fewer resources.

Many pathogens such as cholera, typhoid and e coli will thrive in the human intestine to our detriment if they are able to get there. From their point of view, intestinal life is ideal—regulated warmth, ample moisture and a constantly renewing supply of food. However, the ideal nature of this home means that they multiply with such success that they can overwhelm a host. Killing the host then destroys their perfect home. This means that they must constantly be on the search for new intestines to colonize. Moving from one host to another is their challenge. From our point of view, we want to make this journey impossible for them, so we must do all that we can to isolate our waste—their exit point—from our food sources—their next input. Hence, we need to treat our faeces with such care that even microscopic amounts of these pathogens cannot migrate into our drinking water or our food. The introduction of public sewage systems, putting our waste inside pipes to be taken away from our homes in a watery slurry, was the greatest advance to health that urban humans have ever made.

The challenge of a waste control system is to contain our faeces and then to kill the pathogens. We currently do this by first mixing our stool with clean water as a transport medium, contaminating this water in the process. Then we mix this contaminated slurry with all the grey water coming from our showers, baths, tooth brushing, cooking and washing, making all of it potentially hazardous as well. This large volume of liquid leaves our home, but, along its route to processing, it gets mixed with effluents from every industrial process in the city, making the eventual cleansing much more complex. We create a huge volume, all potentially contaminated, to meet the major design criterion for a treatment plant which is for it to be hidden away in an isolated spot with a fence around it. To connect our homes to this out-of-sight solution, the problem of dealing with pathogens has been transformed into one of transport with the consequent complications of pumps, leakages, blockages and chemical complexity. But it does manage to send our waste out of our lives (good) and out of our sight and minds (not so good). Unfortunately, many people also use this magical out-of-mind disappearing trick for other things that water can carry that they no longer want such as drugs and various types of household chemicals. The treatment process which must be applied to the total volume becomes ever more complicated.

This complexity is completely unnecessary. The bacteria at the heart of the problem are not that difficult to kill. Heat, chlorination, ultraviolet sterilization or microbial filtration can all do the job simply, effectively and relatively cheaply.

If we begin with the remit simply to isolate and sterilize our faeces, then the problem becomes much smaller. The simplest solution is to deal with our waste totally within the family with a composting toilet. We would know exactly where all the pathogens were so we could deal with them without pipes, leaks, or the polluting of some distant neighbourhood. Composting toilets work. I lived with a simple one in a rural area for over a decade with no problem that couldn't be mitigated with a cupful of wood shavings to return the composting process to balance if we had been careless. Annual maintenance and clean out took one morning once a year. Before feeding the result to our roses, we treated with heat to gain further assurance that all pathogens had been killed. If people are put off by the clean out task (not nasty because the composting process had already transformed the waste), then a business of compost specialists could easily be established in the town to install, service, empty and sterilize. Isolating and killing bacteria are not difficult tasks.

This has the bonus of protecting our water. The grey water system becomes isolated from the pathogens, except for negligible amounts which could arise from bathing our bums or washing our underwear. Because of that small possibility, however, grey water should probably not be directly repurposed for drinking. It will not be pure in any event as it also contain soap, and scraps of food, fats and grease from kitchen use. Since excess phosphates from soaps are unwelcome in the marine world, our grey water should also be dealt with locally. This gives us the opportunity to reclaim a valuable resource. The amount of treatment we give to our grey water will depend on the scarcity of water in our world. If water is abundant, our grey water can water our plants or we can treat for phosphates before adding it to storm runoff. In situations where water is scarce, however, we will want to be able to reuse it so more treatment may be needed, the form depending on the eventual use. If we have not flushed it away, reuse will still be an option. Today, we see people living in areas of drought flushing most of their water away just so their waste can be taken out of sight.

It may also be decided to approach this problem jointly within either the neighbourhood or the town. Either of these could lead to quite different options. Consider a few low-rise buildings like Eve's that have formed into a neighbourhood. If they wished to maintain flush toilets, they would, during the renovations, install pipes dedicated specifically to the toilets, thus assuring that faeces can be kept separate from grey water. Coordination between families would be required so that the toilets were located close to the drains which could deliver the slurry to a small treatment system in a basement. A separate drainage system would collect the grey water. Small sewage systems exist commercially now in different sizes, used mainly in remote areas and worksites. They are generally digesters that can treat the resulting product with heat or ultraviolet sterilization. This is essentially a composter located outside the unit, operated by a designated professional. There are a range of options to deal with our faeces that are both local and responsible.

Other waste products, such as drugs or household chemicals, could not find their way into such a system without these actions being challenged by other users. We can demand responsibility locally that is impossible in a large, invisible, impersonal situation. We could then look for ways to deal specifically with each of these unwanted items. Industries that produce harmful effluents will have no general system into which to dump their waste. They will need to neutralize their contamination as part of their production process. Dealing with a problem at its source allows much more targeted solutions.

When the amount of new water that must be introduced into the system drops because we are using less and reusing more, this makes us more secure against dry periods or drought while also reducing the demands we make on common water sources like aquifers. We should not be using water just to transport our waste while also bemoaning that it is becoming an increasingly scarce resource. The simplicity of a local system, with fewer, smaller, more visible pumps and pipes, reduces maintenance costs. And, at the end, we have rich fertilizer for our plants or captured methane for fuel. We can replace some of the artificial fertilizers that are damaging our land and watercourses with excess nitrogen. Our wastes can be transformed into healthy soils and beautiful landscapes. A process that is circular within "ourselves or our doings" has benefits besides the opportunity to be responsible.

Storm sewers to take excess rainfall to the rivers and on to the sea may still be necessary in most areas that have a rainy season. These may require the cooperation of all within a city and so would have to be provided and serviced by a tool. However, this rainwater too can become a resource instead of a problem. Cisterns in each building and town capable of providing for some of our water needs during the rest of the year would allow rainy seasons to make us even more self-sufficient with respect to water. Areas that are vulnerable to extreme drought are increasing. They now include most of the western half of the United States. But these people are still flushing toilets, washing cars, generating green golf courses, often even watering lawns, while loudly demanding that someone else protect them from both drought in the dry season and flooding during the rainy season. Out-of-mind solutions are dangerous in the complacency and lack of personal responsibility they induce.

All of our waste should be handled locally. Sending solid garbage hundreds of miles in search of a landfill is neither responsible nor sustainable. When we must deal with our waste within the area we walk, see and smell, the feedback between our garbage and the problems we face disposing of it is direct. Accumulating waste will force us to demand long-term utility from everything we buy and an obvious life cycle for every product we take into our homes. We will reject packaging that will be a problem to deal with. For the waste we do produce, we can first remove everything organic that can be composted. Since we will already have experts in the community in turning organic material into soil, this can be handled locally. Our local soils will grow thicker and humus-rich. Our roses are going to be even happier.

Next, we must look carefully at what is in our garbage and, where possible, demand either reusable or compostable alternatives. Because many items become garbage simply because they are no longer wanted by some individual, we should provide a spot where such items can find new owners. A "free store" could hold such items—clothing, furniture, appliances and the like—to be perused and claimed before they get moved on. Next, the town could gather and market those items that have a recognized recycling value such as paper, glass, metals, or some plastics, and arrange to deliver them to a tool that will use them as input. If there is a cost to us associated with recycling a particular item, then it should be included in the price of the item and so be borne by the person who initially purchased the item. Such feedback will spur the search for alternatives. For the garbage that remains—those items that cannot be composted, reused or recycled—the town may have to provide safe disposal by fire or burial locally.

Scarcity of space will always be pushing us to reduce the amount that falls into this last category. To encourage this feedback, our garbage must never be hidden away in a distant dump behind a fence. Feedback is powerful but feedback only works with immediate information. Out-of-sight, out-of-mind severs this loop and creates insoluble problems that gradually become accepted as the way things have to be.

Taking personal responsibility for the implications of our food requirements will also cause large changes. Industrial farming practices today create pesticide buildup, nitrogen runoff, reduction in biodiversity, pressure on water and energy systems, and exploited workers. It is all done in answer to our demand for cheap food. We are responsible for the compromises made to feed us. Cooperating in human-scale groups will change the way this works.

In an urban area, there is rarely enough space to grow our own food. There could be fruit or nut trees incorporated into our landscaping and perhaps a few beds of herbs and some vegetables on rooftops or in vertical farming in repurposed warehouses. These possibilities may increase with a greater appreciation for local production, but it is likely that most of our food will have to be sourced outside our town.

One factor in locating many crops today is labour that is available unevenly throughout the year. For vegetables, many hands are needed in the spring for planting, fewer for tilling until harvesting again calls for a crowd with none required during the winter months. For grains and fruit, the demands are highest for a month or two at harvest, being low or non-existent at other times. As farms became more mechanized, many of these tasks were replaced with machines for some crops. Those crops that were difficult to mechanize migrated to parts of the world with impoverished day labourers providing the seasonal workers.

For crops that could be mechanized, land that was too variable for monoculture or too hilly for efficient use of machinery was deemed not useful regardless of its inherent fertility. But where huge, level, irrigated, fertilized fields could be put together, large crops could be generated cheaply and worked solely with machines. These could outcompete all smaller local farms on price. Our dedication to finding the cheapest food forced our smaller local producers to abandon many crops, turning their land to hobby uses, residential subdividing and fallow fields. There is usable land close to most urban centres. Some of it may not yet be paved over.

A connection between a neighbourhood in town and a farm in the adjacent countryside would help both, the town providing a sporadic work force while gaining access to food whose conditions of production are known, the farm gaining the ability to grow crops that require a flexible labour force. The farmers would organize, plan and oversee the work, doing maintenance throughout the year, and perhaps storing the harvest until it was needed. The neighbourhood could have input into varieties, crops and other concerns such as farming methods or fertilizer use, while providing work parties—people doing gift work for their family—when spurts of labour were needed. Together they would be able to produce all the vegetables, fruits and meat that the neighbourhood needed. Production beyond the needs of the neighbourhood would generate cash which could be split between the workers and the farm. There would be no transport damage, no deterioration due to supermarket display, less rejection of food deemed unattractive as we all come to appreciate the compromises necessary to eliminate the last scab from the apple or twist from the carrot. There would be no exploited work force. Consumers of food would have a say in water use, energy use, fertilizer use and pest control. We would gain beautiful, fresh food, lower our living costs and become responsible eaters.

There are, of course, foodstuffs that we would like to have in our lives that cannot be sourced on a local farm. Bananas, oranges, olives, chocolate, coffee and tea cannot be grown in temperate zones, but would all be welcome in my cupboard. Many other manufactured or processed items may also have to be distantly sourced. The town will still need a store where we can buy these items with an assurance that they have been responsibly grown. We are back to responsible consuming again with the need for good information on every item. Connections between that local store and local farmers will produce another route to our sourcing of local food.

Suburban Possibilities

Everything changes with density.

Creating a home of about ten thousand square feet in an area of detached single family dwellings presents a very different architectural challenge than that faced by Eve and her family with their single floor in a low-rise. Most suburban houses contain a few thousand square feet and were designed to house nuclear families of four or five. Now, because of our aging society, they will often hold one or two people who are reluctant to move because of attachment to an area or because the house is full of memories and accumulated stuff. Most bedrooms, second and third bathrooms, and family rooms sit empty except on those rare occasions when someone visits. This space is under-utilized, serving mainly to separate the individuals along the street from each other. To create a connected ten thousand square foot space, it will be necessary to join a few of these houses together, by opening walls in adjacent buildings and by building walkways, courtyards, or infill structures. In areas where houses are spaced significantly far apart on large lots, it may be possible to move a few houses towards each other before joining them with infill buildings or courtyards. Jacking houses up and moving them a short distance can be a valid part of a renovation project. If none of these options are feasible, it may be necessary to take down a few buildings, using the salvaged building materials in a rebuild that suits our needs. Changing our family structure demands appropriate housing. We must not waste good buildings, but we must have structures that reflect our needs.

When such family homes have been created, the neighbourhood will be a dozen or so ten thousand square foot residences ranging over what is now one large or two smaller suburban blocks. A town would contain a few thousand people within a footprint of up to twenty or thirty blocks. This spaciousness changes both our options and our challenges in meeting our energy needs, dealing with our waste and obtaining food.

There is much greater rooftop area on which to locate solar panels allowing solar to contribute more than it could in a high- or low-rise. For many neighbourhoods, this, with battery backup, will be sufficient. For those that need further power, a generator may have to be established in the family, the neighbourhood, the town, or as a tool serving a group of towns. The challenge as the service area increases would include building a grid for delivery. The tradeoffs between efficiency, cost and aesthetics, such as in whether the delivery system should be delivered versus piped hydrogen or surface versus buried electrical wires, would be ours to make.

The considerations are similar when we look at our sewage. It may be most effective to produce and service composters located in each home rather than trying to pipe effluent to a central spot. The town could provide composting expertise, support and service. With more green space, we would have more roses to accept this created soil. The savings in water would mirror the situation in the urban example and would also be welcome in our larger gardens. If, on the other hand, we decide to create a community service with a centrally located digestor, we would have to create and maintain the piping system. This would still exist totally within the range of "ourselves and our doings" so all compromises and feedback would be ours to deal with.

Because the dwellings in a suburban neighbourhood are dispersed, there will not be a spot where the family homes intersect like the ground floor of our high-rise. However, it would still be beneficial to be able to share things like workshops, bike shops, exercise rooms, playrooms, and common lounge and patio space. A building to house these activities that is centrally accessible would be an important neighbourhood amenity. This common space could also house the first level of neighbourhood services such as a power distribution node, and a sorting area for recycling before it is passed on to the town.

There must also be space for a centre in the town. Twenty contiguous blocks in the suburbs now will probably be completely residential, built with the assumption that we will drive to everything we want. Our needs were not expected to be met on foot. Twenty blocks—an area of about four blocks by five—is still very walkable. Activity will return to these streets if more of our lives are located here. On a central block, we should find space to build something larger than a dwelling, with storefronts on the street and offices above, space for shops, a bakery, a food store, the town clinic and bobby, shared offices for remote workers, professional services like dentists, lawyers or accountants, cafés and meeting places. A few thousand people can support these services. Even in the suburbs, we should be able to spend our time in our town, surrounded by individuals we know.

This hub could also hold our transit connection to other parts of the city as well as a parking lot for most of our shared vehicles. A corridor connecting this lot to busier roads at the edges of the town would again remove most cars from our streets. This change would provide even greater benefit here than it did in the urban situation. Since suburban roads have been built around the car, made wide to accommodate parking needs, with extensive paved portions for driveways, laneways and garages, reclaiming these areas of concrete and asphalt, replacing them with grass or all-weather permeable surfaces, could easily double our green space.

As lanes and laneway garages disappear and as fences come down, the interior of each block could become contiguous green space with paths, trees, benches and gazebos connecting our dozen or so homes to each other and to the service building we have constructed. Within this, there could be areas set aside for specific purposes such as a shared play area with swings and slides, a small orchard, a shared herb and salad garden, private quiet spaces or whatever the neighbourhood decides it needs. As we change the nature of our families and the way we spend our days, these houses will have people in them all the time. The common spaces will become similarly populated. Neighbourhood space can be alive.

The space in the front of our houses that we had dedicated to cars offers a different sort of gain. Often lying between neighbourhoods, it can be dedicated to functions less aimed at creating interactive space. Options would include an orchard, a commercial vegetable garden, a small chicken operation to produce eggs in the neighbourhood, a mini-forest or a natural meadow to encourage animals, birds and bees to share our space. These spaces will also enrich our lives as our streets are transformed. We can replace cars with things that are currently absent from our lives.

Our lives will be enhanced by the variety. Research has correlated the benefits of bird song to mental well-being. Living amongst trees cools our world and cleans our air while sucking in carbon. This too has been correlated with enhanced well-being as evidenced in the growth of the "forest bathing" movement. When the town meets some of their own needs for fruit, vegetables and eggs, we support jobs for our neighbours while gaining fresh food whose provenance is known. We can close cycles with direct uses for our grey water, kitchen scraps, and composted waste. Many people initially saw a move to the suburbs as getting closer to nature only to see the aspects that attracted them get turned into parking lots as nature was pushed farther and farther away. The suburbs can be turned into a paradise nested in green space by our ability to cooperate.

Groups of suburban towns will meet some of their needs with regional tools. These could include a regional hospital with specialist doctors, more specialized schools, a larger library, an arena, a larger theatre for travelling drama and concerts, and a setting for those businesses that require a population base in the tens of thousands to be successful. It makes sense to locate these services together. One central town will grow to include larger buildings, more jobs and more people. This regional centre will be the focus for transit from the towns.

Rural Communities

The situation changes again when we turn our attention to rural communities. As the density decreases further, the space between people becomes the dominant feature. How does one begin to form neighbourhoods or towns in this situation?

Sharing the work of running a farm while everyone contributes only a few days of work a week immediately increases the number of people who can live on a farm. As each worker is replaced by three, the number of residents triples. The shift to more labour-intensive crops made possible by the availability of seasonal labour from the city also increases the number of workers who would be needed permanently. Remote work will also allow people to become part of a farm family without being part of the farm enterprise itself. Thus, having a family of ten to fifteen people living and working on a farm will not be difficult. If the homes of these families were located where farms abut, then a small neighbourhood based on three or four farms could be created. These clusters of families may attract others seeking the rural experience but not associated with any of the farms. A group of such neighbourhoods spaced over perhaps one to two hundred farms, though widely spaced, would be their town. At some central point, they could establish stores, café, a meeting place, health-care facilities and schools. This concentration of buildings, people and jobs would become the core of the town. However, the town, ranging miles in every direction, exceeds the range of "ourselves or our doings." Daily lives do not lead to interactions because everyone is either on their land or moving about in a vehicle. Many aspects of this town tend towards becoming tools.

Isolation is presently a defining factor in rural life. This can be ameliorated but not overcome with communication technology. The compensating advantage gained by living in the country is the quiet that comes with spreading people out, the expansive views, easy access to fields, woods and streams, and greater connections with wildness. For some people, these benefits trump urban exuberance. If rural living is based in a cooperative family of ten or more, nested in a small neighbourhood, much of this rural isolation would disappear. Rural living could then become more attractive to people able to work remotely. Farm neighbourhoods might grow. Rural towns could become more vibrant places. The rural experience would become a more attractive option.

The power requirements of machines used to work farms will need to be met sustainably. Open spaces will allow for the unobtrusive siting of banks of solar panels and wind turbines. Running our machines on batteries charged by these sources or by hydrogen that can be generated by them, may be all that is required. But if this is neither sufficient nor reliable enough, then we may need to look at setting up a shared source of power generation. Instead of a grid, it may be most effective to site a modular nuclear reactor in the town core to produce hydrogen as well as local power. This could then be delivered to the farms and used to power the machinery.

Cars are more necessary and harder to share as people spread out. Much of our town will not be accessible on foot. Going to the town centre, the stores, services and meeting places, will require that we drive. On the other hand, a vehicle can easily be parked on a farm without alienating other uses for the space. The roads that connect farms to each other and to the core would be important infrastructure. Transit joining the rural towns to adjacent urban or regional centres may be useful but a dispersed population also offers challenges in the creation of such transit. Park and ride may be the first link in this transit system but, once in their cars, people may very well find it more convenient to drive directly to the regional hub. Once there, transferring to a metro would make the whole urban area accessible.

More space would also offer opportunities in dealing with our waste. Farms have never connected to sewage systems, usually employing a septic digester connected to weeping tiles—essentially an anaerobic composter as an end point of a flush system. With the range of composters now available, dealing responsibly with sewage at the farm level should not be a problem. Farms are in the business of building soils. Composted waste is always an asset. Facilities to help us deal with garbage, such as a free store and a collection and separation facility for recyclables, could be one of the services located in the town core.

We can cooperate to meet all of our needs responsibly within the range of "ourselves or our doings." The feedback within this world makes these solutions sustainable.

CHAPTER FOURTEEN

Looking After Each Other

⌇→

The test of any survival strategy rests in how well it helps us meet our needs. We will now look at some specifics of daily life in this new world where everyone is embedded in a family. What is added to our lives when we live in a neighbourhood of 150 people who share our daily lives? How does a town focus the beliefs of a few thousand people on tasks that benefit us all? Can tools be controlled and still enrich our lives?

When human-scale groups organize our cooperation, we are offered many opportunities to do things for ourselves. This eliminates the isolating experience of relying totally on professionals, strengthens the gift space and reduces our work week. As examples, we will look here at fire protection, health care and education.

Protection from Fire

Fire is an ever-present risk which demands preparation. Human-scale groups are the perfect place to take this on. First, the neighbourhood could organize the purchase of heavy-duty fire extinguishers for every home and hallway.

With standardization and a few evenings of instruction, everyone should be able to wield one effectively. An alarm in any home would summon everyone nearby to come running with the nearest extinguishers. Ten fire extinguishers arriving in thirty seconds would provide a much more powerful and immediate response than evacuation and a ten minute wait for professionals to arrive. The neighbourhood could also arrange an assessment of every home for risks to head off some fires before they occur. Most fires will either be prevented or extinguished immediately.

The town could also organize to replace fire insurance with a pledge that everyone will pitch in, with funds and work, to rebuild any home that is damaged by fire. A few hundred families in a town could share this effort easily. Sharing the risk locally like this would allow people to overcome a fire with a small increase in a few hundred mortgages rather than having the burden borne completely by one unlucky family. It also makes fire risk or risky behaviour the concern of us all: many eyes will root out risk. Only after we have seen what we can accomplish together should we decide to seek out the services of professionals.

If we believe that reliance on an immediate neighbourhood response coupled with a plan for helping to rebuild after a fire is not enough, then the town could look for another level of help. A response at this level may involve buying and storing more powerful extinguishers or hoses in convenient spots and training a more dedicated group of volunteers in their use. These people, likely to be present in the town because of their daily activities, would be able to offer a more powerful two-minute response. Each neighbourhood and each town would come up with different solutions based on the nature of their housing stock, the type of fires they considered likely, the types of resources they had at hand, the closeness of the neighbours, the distance they would be from professional help, if such was supported, and other considerations. Fire protection would be tailored specifically to the needs of each town.

For example, if you live in a rural area and professional fire departments can be expected to take at least ten minutes to arrive, then it would make sense to put your money into small portable devices immediately accessible to residents rather than into a distant group of professionals who can, at best, be expected to arrive in time to save your foundation. Alternatively, if your neighbourhood is a number of floors in a high rise, it would make sense to put your money into hoses in each hallway on a water system with backup power and training volunteers from amongst the people most likely to be in the building. These solutions would both provide people with an immediate, and generally very effective, response for their particular homes.

The group of volunteers could meet regularly to train and plan for the different types of situations they may be likely to face. But, as well as being firefighters, they are also local residents and their training and awareness of risk would become part of their daily life. Sharing their knowledge, formally or informally, would make everyone better prepared. Their presence would make every neighbourhood response to an alarm more organized and more effective.

A fire would immediately trigger an alarm while, if anyone is at home, the fire would be immediately attacked with the home fire extinguisher. Within seconds, neighbours—perhaps including a volunteer or two—would arrive and add their extinguishers to those of the family. The alarm would also initiate a call to each of the volunteer firefighters scattered around town. The neighbours at the fire would contact the gathering volunteers, either cancelling the callout if the neighbourhood response had been effective or giving information to allow the volunteers to be effective when they arrive. Within minutes, the volunteers start to arrive, adding their experience and equipment to the building response. The volunteers would enhance the neighbourhood response. They would not supplant it.

But there are still fires that cannot be handled with neighbourhood and town resources. Even with an immediate response of many fire extinguishers, fires can get out of control. Industrial fires that require specialized knowledge and equipment may be a possibility in some towns. Each town would assess these risks and decide whether they wished to fund a tool to call on in these circumstances. If they decided they wanted a professional service to be available, then they would have to look to adjacent towns to see if there were others who felt the same way. Tools need partners to share the costs.

If enough towns agreed, then a fire hall with professionals could be established, maintained by the group of towns that had agreed to participate. Once again, this tool would be designed to enhance rather than displace local effectiveness. The professionals would be charged with training the volunteers. Once the professionals arrived at a fire, there would already be a crew of volunteers, addressing the situation, ready to support the professional efforts. Everyone would be aware that they are paying for this service, so there would always be an incentive to arrange things so that the drain on professional services is kept to a minimum.

The more towns involved, the more varied the types of expected fires and the types of buildings they will need to service. But some towns may have such specialized needs that they have less need of such a general service. In areas where a town is a cluster of high-rises, the need would be for equipment aimed at dealing with high-rise fires, whether that is evacuation equipment, pre-installed pumps or apparatus to help people safely evacuate smoke-filled buildings. Their choice would be between contributing funds to a general service or spending money on this type of equipment while employing only a few organizers of a larger cadre of volunteers. The nature of the towns that fund them would define the tool.

I have had experience with an informal volunteer system. When I was a teenager in the early sixties, I spent my summers trying to be useful on my uncle's farm in eastern Ontario. The rural phone system was still local, so each farm along the concession road had a phone on the wall with a crank on the side that you could turn to produce a series of long and short rings. These rings were heard in every home, and you were only supposed to pick up and listen if the code you heard was your own. Only when a call was leaving the neighbourhood was an operator engaged. But everyone was alert to another code that was very rare—a long uninterrupted ring that meant immediate help was required by someone on the line. Usually, it meant fire. One evening, as we were just preparing to go to bed and a summer thunderstorm was convulsing outside, there was a particularly bright flash of lightning coupled with a terrific clap of thunder. Less than a minute later, our hearts were stopped by that long ring coming from the telephone.

My uncle, who had just retired for the night, raced out of his bedroom, throwing on clothes as he came. He picked up the receiver, listened briefly without uttering a word to a conversation that now included all the farmers on the road, and then continued his dash out the door grabbing items on his way to the truck. He went racing down the lane at a speed that neither the lane nor the pickup had been designed to handle. The lightning had struck Tommy

Johnson's barn about a mile to the west. The nearest fire truck was about five miles distant and, being manned by volunteers, it would take at least ten to fifteen precious minutes for them to gather and get there. Everyone knew that the fire truck would only be aiming to help protect the house and other outbuildings. The fire in the barn would have had too much of a head start and too much fuel in the stored hay for them to be able to do anything. But the neighbours arrived immediately. They got there in time to rescue all of the stock and to contain the blaze to the barn until the pumper arrived.

By the time my uncle returned the next morning for morning milking, the farmers had already begun arranging amongst themselves to care for the stock until a new barn could be built, to all contribute some hay to get Tommy and his stock through the winter, and to help in the raising of a new barn before snowfall. It was a disaster, sure enough, but no people or animals had been lost and the neighbourhood of fifty or so farmers, with a history of cooperating on many fronts such as in threshing bees and barn raisings, could put it behind them by working together. The planning for this level of fire protection had been almost non-existent, relying as it did on things that they all already knew. The resources—pumps, hoses, axes, bridles, shovels, ropes, telephones—all existed already on each farm, serving other uses. Because it was local, it was fast. It had been the only truly effective response.

The responsibility for protection from fire rests with each individual and family. Anyone may choose to make no provisions, undertake no training, proffer no resources, keep the money, keep their time free from training sessions and trust to luck, fate and good karma. That is a choice. Or they may choose to only pay money to professional firefighters and insurers who promise total security from fire with no personal effort or thought required. Empowerment allows us to choose any solution, but responsibility demands that we live with the consequences of that decision.

Those who want a level of protection that is beyond their own abilities to provide must first look for allies in the neighbourhood willing to share the efforts, the costs and the protection. They will not have to lobby for more tax money to go into firefighting instead of into parks or schools. They will not be forced to pay taxes for fire protection and then hope that staffing decisions made elsewhere will translate into an adequate level of protection for them, their families and their homes. If they have needs, empowered people create solutions.

If we contract for services, the reality must live up to the promises or we can opt out at the next point of renewal. If enough people are dissatisfied, then this particular tool will wither away. The constant possibility of withdrawal of support keeps every tool striving to meet the needs of their clients. Tools will be motivated to evolve into more useful versions of themselves. This is the opposite of today where professional services are so insulated from clients by guaranteed taxation that they soon reflect the needs of the professionals rather than the users.

Enhancing Our Health

Healing relies on informed listening and directed acts of care. Medical technology provides marvellous tools to help us, but healing is, at its heart, a personal transaction. Diseases are relieved with time, care, comfort and chicken soup as well as with the fruits of tools like drugs, surgery, radiation or dialysis. Healing is an act of human connection.

The more personal our care, the more bearable are bouts of ill health, the more effective the course of treatment.

Our first support in illness will be found within our family. Ten to fifteen people, with a variety of ages, skills and schedules, will provide options for care in the home that do not currently exist. Sharing this load will permit comfortable bed rest, the most necessary contribution to recuperation in most cases. When this family support is nested within a neighbourhood of friends who are also offering to help, the care will not be as onerous or impossible as it would be for any one individual. Neighbours know that such acts of support will be reciprocated whenever the shoe is on the other foot.

Emergencies such as wound dressing, dealing with fever, stabilization after a fall and a need for CPR will be addressed safely and quickly because every member of every family will undergo first aid training as a basic life skill. Today, a person can bleed to death, die from a heart attack, or suffer permanent disability from being moved before being stabilized after a fall, even though people were present, all impotently standing by, waiting for professionals to arrive. Such widely spread first aid knowledge will make everyone insist that their home have a basic kit with dressings, splints, hot and cold packs, antiseptics and a small group of generic drugs. This training, probably arranged through the town or the school, and with refreshers requiring only the occasional evening, would be a significant contribution to many positive outcomes. The cost in both time and money would be small.

We suffer from a case of what the social critic Ivan Illich called the "disabling professions." Once we have defined an area as being the responsibility of a professional, be it firefighting or doctoring, we assume that, without years of specialized training, we are so inadequate as to be helpless—we make ourselves disabled. We deem all activity in this area as inherently beyond us even though basic competence would yield important gains.

The family is also the proper venue for rooting out sources of ill health and introducing healthy habits. Our families know us best and it is they who will have the burden of dealing with our ill health or the joy of our robust aging in the future. Family members will be motivated to make future illnesses less likely.

For example, stress is bad for the human body. The competitive nature of work today means that jobs often induce stress when we have them and lead to even more stress when we do not. Trying to live a balanced life in the face of ever-increasing work demands is also a recipe for stress. A two-day work week with an assurance that our community will help us find an appropriate job should the need arise will lift that cloud. Also, isolation is widespread today as the size of the family trends downwards. This can induce depression, which triggers other ills. In a family, everyone is surrounded by people who have their back. This too will reduce stress. However, if an individual is feeling stressed, then the family will see it. Knowing him, they can identify the stressors, understand where they come from, and make them go away.

Dependencies on cigarettes, alcohol or other drugs can create conditions that help diseases to flourish. If, as a family or neighbourhood or town, we expect to be asked to contribute care for someone with these diseases in the future, then we have a stake in today's decisions that create the risk. Currently, lifestyle decisions are considered personal choices while the care for illnesses they generate falls to society. Being responsible for ourselves means that decisions are ours to make but the consequences of these decisions must also be ours. If an individual's family is willing to commit the time, resources and heartbreak necessary to deal with lung cancer, heart disease or liver failure, regardless of present activity, then this lack of responsibility to the group can be coddled. However, it is more likely that they will turn their health-care efforts towards helping him to avoid the disease. If someone flat out refuses to change his habits, then there may have to be limits placed on the aid available when these diseases do occur.

Of course, compromises are still available. Someone who insists on smoking could be asked to contribute resources over their lifetime to fund their more probable future care. Actuaries can turn the risks of smoking into an appropriate fee. This would allow an individual to continue the habit without burdening others with the consequences. If the premium demanded was greater than an individual could realistically bear, or if he declined this option, then these are facts that both the smoker and the family should know as they move forward.

If quitting is the choice, then local support will be a significant help. When everyone in the neighbourhood is aware of what is being attempted, then a zone of support could instantly appear. Temptations would be shooed away. The person trying to quit would be motivated to gain his neighbour's approval. The most successful programs in dealing with addictive behaviours are those, such as AA, that mobilize a supportive group of individuals that cares that the person succeed.

The family is also a natural setting for making exercise an integral part of everyone's life. Exercise is easier if we have someone to remind us, to share our walks, workouts, cycling and games. The neighbourhood and the town can also produce facilities to encourage fitness. Shared weight or circuit training equipment could be situated in the

14. Looking After Each Other

neighbourhood. Parks with soccer pitches and baseball diamonds could become uses for emerging green spaces in the town. Some venues such as basketball courts, indoor pools or skating rinks may have to rely on a tool serving groups of towns or the city. If physical activity is valued, then facilities to make it a seamless part of our days can be created. This appreciation for activity starts in the family.

Pollutants in our homes, in the food we eat, in the air we breathe, and in the ground around us, must be cleaned up so we will not have to deal with the health consequences later. Even if much pollution cannot be rectified in the family, it is here that the problem is felt because it is here that people get sick. Families must be the ones that name the problem. Only then can a search for solutions begin. The alternative is to passively accept these threats as a cost of our wealth. We must connect the potential illness to the actions that created the risk and then identify those standing behind those actions, whether they are individuals, industrial tools or the people demanding their service. We must demand that they be responsible for the harm they add to our lives. This will start by publicizing the damage and looking for allies. This may advance to a boycott. It may even mean millions of individuals changing their behaviour to rectify damage to which we are all contributors.

Addressing health risks like these is never a government problem, no matter how large a tool we must create to implement a solution. When the problem is our personal health degradation, then every action, even when coordinated with those of others, is done to improve our own personal health. Each tool we create is funded with our personal resources. Action starts and results end with the individual.

After the family has looked at destructive personal habits, stress, nutrition, exercise, first aid preparedness, and health hazards in the environment, we could turn to the neighbourhood and town if we want further healing options. Sponsoring a few people who spend much of their time in the neighbourhood to take a more detailed first aid course would improve our preparedness for accidents. Such courses, lasting a few weeks and focussing on stabilizing serious medical emergencies for transfer, are now available to first aid workers on remote worksites. Having this skill scattered throughout the town would save lives by reducing the crucial time between emergency and knowledgeable care.

We would likely also want access to sophisticated knowledge for the care of serious problems. We may decide to provide resources for a doctor located in a clinic in the town. This doctor, as well as staffing a clinic, would support and help train the volunteer responders. She would be our point of connection to even more sophisticated tools.

Medical technology can do astounding things. If possible, we would want access to these capabilities. A group of towns could support a small hospital with recovery beds, labs, radiology, rehab services, birthing rooms, an emergency room, outpatient support, day surgeries, dialysis and other treatments that can be delivered close to the community. Specialists could be accessed here, either in person or remotely. These small hospitals would be the link between larger, more distant tools and the community, allowing hospital care to be delivered close to our circles of support.

Not all the equipment and knowledge we wish to have available could be sited in local hospitals. We would benefit from larger hospitals with sophisticated labs, a range of resident specialists and the ability to offer intensive care. This facility would support the small hospitals in its area, making the care they deliver more effective. Seriously ill patients would be in the large hospitals; most care would occur in the smaller ones.

Small and large hospitals would work together to provide a continuous range of service. For example, surgery could be done in the regional hospital with a recuperative period in a local bed. Drug dispensing, lab services, sterilization and diagnostic advice could be centrally located with connections by Internet and a dedicated delivery service making these services feel local. Keeping people close to their family, but with sophisticated services accessible when needed, would be the goal.

Beyond these regional hospitals there are even more specialized services, crucially important but rarely needed, that we may want to have available. We could choose to fund tools specifically tailored to a particular type of illness. For example, there could be a cancer centre, a paediatric hospital, a rehab centre for spinal cord injuries and so on. Empowering the individual does not mean eschewing sophistication. Instead, it means making sure that the tool is focussed on the individual it serves.

In these specialized tools, family caregivers should still have a place. Where possible, pairs of rooms, one for a patient and one for a caregiver from the family, could strive to retain the human aspect of healing. Medical staff could concentrate on the medical aspects of care while leaving feeding, cleaning, entertaining and encouragement to people from the family and the neighbourhood.

Education

Helping our children become competent, confident, contributing adults is the main task with which every generation is entrusted. Success produces both personal and societal benefits; failure guarantees both unhappiness and social collapse. We need settings and tools that focus exclusively on what is best for our children. Children currently must find their way in a world designed to meet the needs of the machine. They are simply another problem that has to be solved. First and foremost, childcare must be such that it minimizes interference with our busy jobs and crowded lives. As they make their way to adulthood, children must navigate largely absent, overworked parents, a distant and absent extended family, an unfiltered Internet for mis-education, peer pressures amplified by early age stratification in schools, organized competitive fun, and schooling that can often resemble babysitting. Some of them shrug these stresses off and grow up fine; others are less successful. A human-scale world of families, neighbourhoods and towns offers us a different range of possibilities.

During the first few years of every baby's life, she needs someone to mediate between her and the rest of her world. For a baby, everything is new and potentially threatening, so someone must provide security, keeping the world at bay while focussing completely on her needs. The constant availability of a principal caregiver during this period of helplessness has been a feature of every successful society.

The first cry of a newborn can be calmed with food, touch, warmth, a beating heart and a soothing voice. The person who provides these things becomes the baby's world. Whenever the baby is fearful or hungry, this caregiver is loudly demanded, initially with a ferocity that expresses her fears of abandonment, but with less and less desperation as she becomes convinced that someone will answer her cry.

This caregiver represents food, warmth, protection, security and entertainment. The rest of the noisy world is of no consequence as long as the caregiver is nearby. As the child gains skills and confidence, she notices and begins to interact with the rest of the world. During these first few years, she journeys from total dependence to wanting to follow her curiosity into independent exploration.

If this task of primary caregiver must be shifted from one person to another, then a new bond of trust must be established. The baby must once again be assured that she lives in a secure world. This may be difficult because the disappearance of a first caregiver only validates her primal fears of abandonment. Permanence, consistency and constant availability are crucial during this initial period.

When she fully trusts that her needs will be met when she cries, she will start to allow the primary caregiver out of her sight. This may take time. Until then, the job of primary caregiver can be a consuming task. However, it must be done for as long as it takes. Failing our children here will result in persistent insecurities; the social fabric such individuals weave will be tattered. The tenor of our society, many of our social problems, all of our hopes for the future, will grow out of our successes or our failures in these first few years. Secure attachment is crucial for the rest of human development.

No individual can do this job by themselves. In no troop, tribe or village has one person been expected to take on this calling without a substantial network of support. Where support for the primary caregiver is weak, the job becomes impossible, compromises become unavoidable, and the task will be poorly done. No isolated caregiver can meet her own needs adequately while responding to the constant demands of a child. A varied family of ten to fifteen people nested in a supportive neighbourhood is able to give the care of newborns the importance it deserves. While the demands on the primary caregiver are still continuous, others must take on all of her other responsibilities.

All regular chores such as cooking, cleaning, gardening, wage work, being a medical volunteer or firefighter, and whatever other projects she was engaged in, must be picked up by others in the group. This is a part of their contribution to producing a healthy child. Raising a child really does take a village. We must all be involved in raising the next generation.

Other family members will gradually become familiar and accepted by the child as surrogates for the primary caregiver. At that point, these others can start to take on some aspects of childcare while the primary caregiver can reclaim some of the other parts of her life. Even without language, children can clearly indicate their acceptance of new people, or they can reiterate their demand for their one special person. Some tasks and some times may still require the primary caregiver while, at other times and for other tasks, other family members will start to be accepted.

We are a curious species. As the baby grows into a toddler, she will feel that urge to explore her world. She will start to find the variety that exists within the family stimulating instead of overpowering. Soon, even this setting will become sufficiently predictable that she will want to poke her nose into the neighbourhood, a larger and scarier place where retreat into the safety of the family is always available. All 150 neighbours will become familiar to her while, in their turn, they will all come to know her as they look out for her. Here, she will find playmates, babysitters and a wider variety of people and experiences. This exploration will be her education as she continues to mature.

At some point, the neighbourhood too will become predictable, and her growing confidence will spur her towards new experiences. For these four-, five- or six-year-olds, the town could provide parks with grassy areas, slides, teeter-totters, sand boxes and room to run and climb safely,

14. Looking After Each Other

all designed to encourage unstructured group play. There could also be some slightly more structured offerings such as dance, swimming or gymnastics. Her world will unfold at a pace that she chooses.

A transition from home to school should be a response to a child wanting larger horizons. She should venture into this new range of experiences when she is ready rather than at a specifically mandated age. Supervised spots for children to gather for play without an accompanying family member would attract four-, five- and six-year-olds. These places would try to make themselves attractive to curious minds rather than being tasked with offering a mandated curriculum. As they would include other children, stories, shared snacks, art and games, music and play, most children would take to this. If a child does not, then the caregivers and the teacher should look at the reasons and see if there are changes on either end that might help. Perhaps an interest particular to the child could be incorporated; perhaps the grouping of children is too exuberant, or some individuals are too dominant, and more safety needs to be built into the structure. If none of that makes attendance attractive, then the child could remain within the family and the neighbourhood where they are comfortable until they decide to join the scheduled gatherings. If many children are not attracted to the local school, then the shortcoming is most likely to be in the school and that should be corrected. Children could start attending such a play school as soon as they wanted to and as frequently as they wished.

One possible arrangement could have a room central to a few neighbourhoods with a dedicated teacher aided by a roster of volunteers as required. In three neighbourhoods there would be about 400 people so we could expect a dozen or so kids in that age range. If there are fewer kids, then more neighbourhoods would have to cooperate. If there are more kids then it may be useful to establish these first schools in fewer neighbourhoods. This will keep the group of kids to a size large enough to encourage interactions and allow coalitions of friends to form, but small enough so that everything can easily be familiar and personal.

By the ages of seven or eight, and perhaps even before, most kids would be seeking a larger scope for new experiences. There is also benefit to the community for children to start mastering specific skills at this age. The town could establish a school with three or four classrooms for children up to the age of twelve or thirteen, designed to challenge curious minds and active bodies as well as to impart specific competencies. Everyone would be expected to attain basic levels of literacy and numeracy, as well as familiarity with basic living skills such as money-handling, nutrition, horticulture, first aid, firefighting and self-defense. But mainly, the mandate of these schools would be to help young people follow their curiosity in whatever directions their questions may take them. As this school would still be embedded in the town, the pupils would already be familiar with their teachers, their classmates and any volunteers. In a town of a few thousand people, such a school may expect to have about fifty to one hundred pupils. If a child is not attracted to the school, they could still easily master the basic competencies within the family.

Young adults from the ages of about twelve to eighteen would benefit from a greater variety of courses and experiences necessitating a larger school serving an area larger than the town. This is the point where the school becomes a tool. It is important to note that the user of the tool is the student, not the parent or the teacher. Movement to this institution would be driven by curiosity as young people answer the urge to expand their social and educational

14. Looking After Each Other

experiences. Groups of students sharing an interest in things like geometry, creative writing, geology, auto mechanics, Spanish or basketball should be able to use this tool to further their explorations. The tool must be constructed to be responsive to the curiosity of the learners.

Here, the community can also make other expectations known to them. During this period, each person should learn a trade, gaining a skill that they could use, either to make a living or to contribute voluntarily to the family, neighbourhood and town. The skills of carpenter, medic, plumber, electrician, mechanic, welder, cook and gardener, as well as many others, would spread through the population. Every neighbourhood would become more self-reliant. For those who wish to follow a trade as a career, mentorships would be arranged, with apprenticeships either replacing school for a year or two or integrated into the school terms. Helping the next generation to become useful functioning adults is a task for everyone. By school leaving, every person should have a useful marketable trade.

Everyone should also acquire the ability to be creative in at least one discipline—drawing and painting, pottery, woodworking, jewellery making, sculpture, metal fabrication, music, words or dance. When a student leaves school, they should also have advanced their basic skills in literacy, numeracy, critical thinking, and self-defence to the point where they can fully contribute to their community. They will be expected to have the basic knowledge of history, geography, science and technology necessary to participate in community discussions. They will be ready to be adults in society.

There will also be a need for tools that impart specific training—medical schools, dental schools, technical schools and the like. The smaller the number of people that want to study in a particular field, the larger the area that will have to cooperate in establishing the tool to do this.

There is also a need for a tool that will engage our curiosity after we join the adult world. These should be places of ideas and discussion accessible to everyone. Wage work commitments would be short enough that following curiosity would be important in everyone's life. These tools would include libraries, access to individuals with specialized knowledge who could become mentors, facilities for research, structured courses of instruction, informal discussions and formal events. This tool should aim to become an important hub in many lives.

An educated and critical populace is essential if freedom is to flourish. We must all be able to express ourselves, to understand others, to discuss critically and respectfully and to forge intelligent compromises. We must all be able to read critically and write clearly. We must all be able to engage with a range of ideas against which we can continuously test and refine our beliefs. If we wish to be free, we must value critical thinking. This tool for lifelong learning will be crucial to our freedom.

Thus, education would involve a gradual widening of horizons from those first months where our world extends no farther than our primary caregiver until that moment when we have taken our place as an adult member of our community.

Innovation

We are the innovator. The most certain thing about our future is that it will involve change, usually beginning with technology and flowing from there into other aspects of our lives. We live in the most technologically fertile period of human existence. In the last century, our world has transformed itself. In the next, it will do no less.

We will see new tools of which we cannot even dream today, both enhancing our lives and tossing up unexpected challenges. As our possibilities change, we must be able to revise our societies. In the face of this change, we must be able to preserve the values on which our communities rest. We must continue to treasure equality, freedom, responsibility and a world based on our ability to cooperate.

Some new options will offer opportunities for individuals to gain at the expense of the community. The Dominator will never be banished. He can only be constrained and the ways we check his behaviour will have to be subject to change. We must always be on guard for threats to equality lurking within these emerging changes. We must always make sure that we are the users deploying the technical wonders, not their servants.

The communication revolution that is still unfolding can bring all the information in the world onto our desks. We have tools that can pluck a desired fact from a mass of data whenever we ask. We now take this constant access to knowledge for granted, overlooking the revolutionary change that it embodies. Information has always conferred power. Universal access to knowledge carries the possibility of levelling many hierarchies. But this possibility often shimmers out of reach because the tool has been sold to the highest bidder. The user of the tool is anyone with enough money to rent space on our screens. Many of these users have reasons to substitute lies for facts. Some are foreign states who wish us harm; others are people peddling hate in hopes of benefitting from the resulting divisions and conflict. Just as information can promise power, bad information can nullify our abilities to act responsibly.

The most recent iterations in information technology go beyond making information available when we ask. Web sites have created algorithms to note our searches, the things we like, the things we purchase, the people with whom we connect and how long our eyeballs dwell on any particular part of the screen. With this data, they create a picture of our interests and our psychological makeup which they refine with every click of our mouse. Advertisers pay handsomely for these details. To raise the value of ad space, site managers must keep us on their page so they push emotive content that mirrors this picture of us they have created while rejecting everything that does not fit. This has extended as far as curating the news a person sees so that only those items which justify a particular world view or inflame passions against opposing views are presented. People are confined to their silo, seeing that portion of reality that corresponds with prior prejudices. This manipulation is done invisibly and anonymously. These tools are getting more efficient at controlling the content of our minds. Demagogues have taken notice. We must assert ourselves and become the users of this tool before it is too late.

These dangers were unintended. In most cases, idealistic innovators wanted to deliver the positive benefits they envisioned. To that end, they created a tool. That tool worked best if there was no central controller, if power was spread throughout, if content could be produced anywhere and if the flow of content was ordered by an algorithmic network capable of refining itself. They saw advertisers as an evil, necessary to pay for the construction of the marvellous tool. But it was still very much a tool, capable of good or ill, morally neutral until the direction was supplied by a user. Our contribution was a demand for cheapness: we demanded that everything appear to be free. And those who were willing to pay started calling the shots. In the process, we became the product that was being sold. The battle for control of this tool, and others that are yet to appear, will go a long way in shaping our future. If we want this tool to serve our needs, we will need to pay for it. We can yet become the user of this tool.

A tool is just a possibility. Structuring a tool so that it is responsive to the user requires both thoughtful design and a user's willingness to pay to support it. Tools cannot be allowed to be subverted by the Dominator as he will use them to expand his control. A tool sufficiently powerful to accomplish great good is sufficiently powerful to do great harm. We must fight for the ability to act responsibly in each possibility that change presents.

Future Demographic Effects

The loosening of the demands of the workplace will allow people to live where they wish. Significant migration could be the result—perhaps from the city to the rural areas or from one part of the world to another. In the past, large movements of people occurred mainly because they were driven, either by hunger or by political upheaval. Enslaved peoples were forcibly relocated to service plantation societies. Peasants were pushed into slums to meet the needs of industrial development. People fled from the slums to the suburbs as the modern world oriented itself around the automobile. Migrations always triggered massive political, social and economic changes. The transformation triggered by the one- to two-day work week will do no less. We can be sure that the picture we have been outlining will be no more than a first draft. Billions of free choices as people seek their best lives in these new circumstances will redefine our future in unexpected ways.

Birthrate may be affected by this change. As the responsibility for raising a child comes to be shared by families and neighbours, the financial and psychological burdens that now fall exclusively on parents would be lessened. Sharing work in teams will mean that choosing to have a child will no longer threaten to scupper career prospects. One reason for the current falling birthrate is the impossible physical, psychological and economic cost involved in the choice to have children. In a cooperative world, more people might well choose to have children.

This would have serious implications. We currently have enough food in the world to feed all eight billion of us (as of January 2023) and we have the capacity to increase this considerably. Much food is wasted—some of this waste is inevitable, but there are large gains to be had in refrigeration and storage and transportation infrastructure. Areas with the impoverished workers essential for many labour intensive crops generally also have underdeveloped infrastructure. Some food choices such as beef are inherently inefficient as they use excessive amounts of land and water for the protein produced, so a movement to plant-based alternatives—or even to more efficient meats like chicken—would feed more people. Connections between farm and city will also allow the cropping of much fertile but currently underused land. There are also possibilities to engineer nutritious foodstuffs in a more industrial model. The world currently can produce more food.

Countering these opportunities will be weather events brought on by climate change and the damage to land and sea that we have already caused. Harvests may become more uncertain. We will have to be adaptable to keep everyone fed but we will be more successful if we can approach these risks cooperatively, sharing ideas, technology, innovation, financing and belt-tightening. Food is at the core of all survival strategies. A dearth stimulates the urge to hoard and lets the Dominator escape his constraints. We can now feed our numbers and we seem to have the abilities to react so that this will continue to be so. When we approach the carrying capacity of our land, as we must do at some point if fertility rates go up, our reaction must be to drop the fertility rate to fit the number of mouths to the available food. Cooperation can allow world population to stabilize at a point below our ability to produce food. This would be the ultimate test for global decision-making through our system of active citizenship.

But fertility rates may not rise significantly. Everywhere where free choice has been assured, small families have become the norm. Where large families persist, it is because of poverty, lack of education and autocratic domination by men. These factors can all be opposed with transfers until they are no longer determinative. It seems that the likely course is that world population will peak soon and then begin to decline. This will usher in a world where food is abundant. There will also be more houses, more cars, more furniture and more of everything than will be required so most production can decrease drastically. We can turn our efforts towards reclaiming resources embedded in our excess items making resource extraction from the commons largely unnecessary. Sustainable power sources will drive the world allowing coal, oil and gas to stay in the ground.

As large families establish themselves in tight neighbourhoods, population density will increase. People will in-migrate from isolated inefficient living spaces elsewhere to join these dense neighbourhoods. This will be true in urban settings, in suburban towns and in rural towns even as the total population drops. We could expect a few things to happen. First, green spaces will appear between urban and suburban towns. These will come to include orchards, vegetables patches, egg and chicken production, meadows and similar uses that do not require vast fields. There will be more room for recreational resources like bike paths or soccer fields. In rural areas, more people on farms working good land more intensively will allow marginal land to return to forest benefitting birds, animals, the climate and the planet.

As wage work becomes less important in our lives, we will reorient towards what we want to do rather than what we have to do. The world will adjust to accommodate the innovator, the explorer, the artist, the creator and the problem solver in everyone, not just the lucky few. Our societies will be defined more and more by the enthusiasms of human curiosity.

The historian Yuval Noah Harari, in his book *Homo Deus: A Brief History of Tomorrow*, starts with the statement that, though our past has been defined by war, plague and famine, our future will be guided by our desires for bliss, immortality and divinity. He then proceeds to look at what that may look like.[1] We are on the verge of great change. When we let the Cooperator shape our future, anything is possible.

The psychologist Abraham Maslow, in laying out his hierarchy of human needs, divides them into five categories. On the first level are our basic needs for food, water, warmth and rest. We are a rich enough world to make this available to everyone. The second need is for security and safety. This can be guaranteed if we dedicate ourselves to equality. The third level contains psychological needs, the first of which he terms belongingness and love. We are creatures who need to be embedded in nurturing families and neighbourhoods. We need only choose to produce them. He describes our need for esteem, prestige and the recognition of our fellows as the fourth level. For this to be real, it must arise from the creation of a true reputation that can only flow from regular contact in a world of "ourselves or our doings." These first four categories of human needs will flow naturally from the decision to base our lives on cooperation. We can then turn our strivings towards the fifth category of human need, that which Maslow has called self-fulfillment or self-actualization, however each of us chooses to define this. Creating a world where the first four levels of human needs will be met for everyone will launch humanity into this new realm.[2]

The bloom of self-fulfillment will look different for every person. However, a world where billions are striving for self-fulfillment will be very different from one where the exertions of billions are driven by their need to obtain food, shelter, security, belongingness and recognition. This future will be a revelation. A golden age awaits.

The worm in the apple is the Dominator, lurking within each of us, ready to assert himself and drive us back into a dog-eat-dog world. The power of this urge must never be underestimated. We must devise strong controls to protect our possibilities. This is the focus of the next chapter.

CHAPTER FIFTEEN

The Dominator

∿➔

Everyone will continue to feel the urge to dominate, to win by making others lose, to show ourselves stronger than others in order to take from them what they have. This drive is intrinsic to the human animal. It cannot be banished with an argument or a choice. We will need structures to control this part of our nature.

We found power in groups when we learned to stand together to expel the alphas from our midst. This stance became codified into the taboos that controlled the drive to dominate. If bullying did erupt despite the taboos, a coalition of the rest always stood ready to oppose the offender. But the Dominator was not eliminated. He was just put in chains.

Ten thousand years ago, some tribes turned from hunting and gathering to cultivation, producing their food in a periodic harvest which could be eaten throughout the year. Fixed fields and granaries became so critical to survival that a cadre of soldiers was empowered as defenders. These powerful men could no longer be constrained by coalitions of their fellows. The old taboos became unenforceable. The Dominator was free. Answering this drive, these soldiers impelled their societies into wars, exploitation, slavery, appropriation and genocide. The hierarchies of feudalism defined our world.

The urge to dominate is too basic to humans and too insistent to be thwarted by shame, religion, conscience or argument alone, even though these constructs may all play a part. We need explicit enforceable taboos to exert control lest we drift into an acceptance of activities that undermine our cooperation and foster inequality. We must all agree that certain activities are beyond the pale and that it is the duty of every person to name them and resist them.

Bullies cannot prosper where people are prepared to name a breaching of the taboos. Only when everyone is assured that proscribed acts will be opposed will we be safe from attempts to dominate us. And when we are assured that our dominating actions will be opposed, we will control our own urges. Taboos on domination demand that we make the effort to consider others, to be responsible.

This may sound onerous or a bit like vigilantism but it will be less so in practice. When people have agreed that irresponsible acts are unacceptable and will be opposed, then norms that reflect that will direct behaviour. In tribal times, in classical Athens, in the medieval city-states and in the independent towns in early America, those who acted responsibly came to be admired while those who didn't were shunned. Every act contributed to a person's reputation and your reputation was of great importance. These expectations were what shaped behaviour.

When an irresponsible act is challenged, the response can involve explanations about the context of that action, and the perpetrator's specific needs and understanding of the situation. They will be faced with the specific pain, specific damages and specific anger that has resulted from their actions. Nothing is hidden in the abstract. Apologies can be offered; restitution can be arranged. Everyone can learn. An incident can be dealt with and put into the past. If harm persists in the face of community sanction, then individuals are risking their place in the community.

The ability to form effective coalitions relies on equality. A disparity of wealth or power allows winning to become an accepted fact because the more powerful can always press their advantage. Small wins increase the power imbalance and inequality grows. The more powerful can grow to become untouchable. They will be able to dominate with impunity.

What does all this mean for us in a human-scale world?

First, we must all feel safe in our family. Throughout feudal times, both church and state have encouraged the nuclear family to be dominated by a patriarch. His control was bolstered by his greater physical strength, a monopoly on the right to earn, church dogma and legal protections. He could beat his wife and children, order their lives in ways that caused distress, grant or withhold necessities such as food, and no one in the community could rise to their defence. This has been challenged but still lurks in our habitual responses. Laws have been changed but the residue of past inequalities remain. Women have gained the power that comes with the right to earn.

Family groups have become smaller. If bullying does arise, individuals must demand equality standing alone. The possibility for effective coalitions has disappeared. However, a family of ten to fifteen people comprised of a variety of adults, children and elders, will permit the coalitions that can name and oppose dominating acts. Fathers and mothers with children can still exist within some family units, but the encompassing group will be large enough and diverse enough to deny the right to bully. Families must be clear that they will oppose bullying.

Being a part of such a coalition carries an implicit offer to help in opposing attempts to intimidate. A few members must be able to stand against the strongest. A group with a great disparity between the physically weaker and the strongest members will offer the chance to dominate. Physical abilities should be spread through the group.

There are forms of martial arts that allow technique, attitude and conditioning to compensate in some measure for differences in size and strength. Everyone should be expected to learn and practice one of these as a basic life skill. If everyone can stand up for themselves, then that alone can discourage the urge to dominate. Coalitions will be more effective. Two or three trained people can say no to larger, stronger bullies. If we do not have such training as an expectation of citizenship, we must invite professionals—soldiers and policemen—into our lives to oppose the bullies we identify. That introduces a force into our midst of which we are unable to demand responsibility with coalitions of citizens. As in feudal times, those who control this force can dominate society. A family that is numerous, varied, trained and adamant about equality is the only guarantee that bullying will be impossible in our homes. Everyone demanding equality will make our families safe.

Amongst the 150 or so people living in the neighbourhood, all known to each other by character and reputation, a general agreement that certain acts are unacceptable and will be opposed can make them disappear. No one would want their reputation sullied in this intimate arena where memories can linger after the act has been dealt with. And if it becomes necessary to name abuse and confront a transgressor, that can be done most effectively where we know the act, the actor, and the victim and have access to a range of possibilities for restitution and closure.

If members of a town agree that violence or bullying are unacceptable ways to resolve conflict, then they must create other mechanisms to deal with differences. This may involve making mediators available for those times when differences seem unbridgeable or tempers reach a boiling point. It could take the form of a prior agreement to accept the wisdom of a designated arbiter or group of local elders with everyone committing to abide by their solution. Doing away with conflict also involves a commitment to support all parties as they work to get beyond a dispute. This may involve obtaining counselling for individuals who are acting out past traumas. It may involve helping all members of a family into new situations. We must do whatever is necessary. Banishing violence demands that we create avenues to get beyond disagreements.

If everyone knows that their neighbours are prepared to come to their aid as they themselves are prepared to go to the aid of everyone else in the neighbourhood, then aggressive acts become impossible. A street after dark is not scary if you know that a call of distress will cause lights to snap on and doors to fly open bringing allies streaming to our aid. Those hoping to prey on others will be aware that our neighbourhood is not made up of isolated individuals. We can take back the streets, yards, alleys and parks when we all accept that we are responsible for their safety. The delusion that more policemen, tougher sentences and public apathy is a route to security is a bill of goods pushed by Dominators who want citizens to be isolated and powerless.

In the town, where reputations are more superficial, coalitions against obvious eruptions of violence would still be able to form in response to a breach of a stated taboo. But long-term solutions would have to involve an individual's neighbourhood where people are more intimately known.

Our shared belief that irresponsibility is wrong will define the unacceptable behaviours. These norms will regulate behaviour. Acts outside of these norms must garner a response from all in the area.

People in a town may decide to empower one of their own to take the lead in security matters. A local bobby with an office in the town core could be at the centre of any response. She would support coalitions within the family or neighbourhood, offering the neutral perspective of a trained mediator. She could be a bridge to other services such as other types of mediation or mental health support. Answerable to the town and funded by the citizens in it, her broad remit would be to help people make themselves secure. Her actions would range well beyond enforcing a code into anything involved with safety such as coordinating responses to natural disasters or ensuring that traffic movements are safe for children.

To deal with activities where effects are spread over large areas rather than localized within families, neighbourhoods or towns, we would need to establish tools. These could include services impractical to establish in a town such as disaster response, forensics labs or specialized investigators. Other tools would be necessary to coordinate situations that involve more than one town.

The urge to cooperate with others with whom we share space, reputation, experiences, history, goals, stories, resources and customs is natural to us. This is our tribe. But towards strangers, the instinctive human reaction is suspicion, which can easily escalate to hostility. No reputation is attached to a stranger so we do not know at a glance who is a taker, who might be generous, who is a helper and who might be a threat. This human tendency to othering has two important implications for our security.

First, there must be tools designed specifically to reduce the difference of the "other." The more we know about others, the more our perceived similarities make the differences seem less important. "Others" can be seen as different in superficial features rather than as the face of a potential enemy. This emphasizes the need for active civic bodies covering large areas, for projects shared with distant partners, for exchange visits to get to know "others" as individuals, for international transfers and for global equality. We must be able to see our common concerns, problems, hopes and fears. Civic bodies, with trappings of shared identity, with common rituals of civic engagement, and with common projects to which all contribute, are able to tie disparate peoples together. The allegiance of widely dispersed citizens to huge states today proves that people can feel themselves to be part of a constructed civic body shared with strangers. The world is already a highly connected place. Technology can help make us aware that we are one.

We do not need an increase in the number of our civic bodies to do this. We just need organizations that will lead us towards universal connection, as opposed to defining themselves around lines on a map whose purpose has always been to separate "us" from "them." Larger than our city, we may need watershed or regional governments to coordinate actions within ecologically coherent areas. Above them, we should only need a worldwide organization to express our common humanity.

Our current attempt at a world council, the United Nations, deliberately undercuts effective coalitions of middling powers in both of its chambers. The security council gives vetos to a group of alpha states, assuring that bullies will get to rule the chamber in perpetuity. The general assembly, on the other hand, gives every tiny state an equal voice, allowing the powerless to overwhelm the middling powers, turning the chamber into an impotent venue of posturing and virtue signalling. Instead, we need a representative body of equals, connected to the people it represents, small enough that reputations must count and large enough that useful coalitions can form. Such a body could actually achieve the goals that the UN trumpets. Security should never rest on the strongest in the room, the policeman of the world. That only assures an attempt at domination. Only effective coalitions of the middling can promote equality.

The need to reduce the fear associated with "the other" also reinforces the importance of transfers at all levels, but especially between distant and unequal regions. Persisting inequalities are the most powerful creator of othering, and this is a lingering threat to us all. As stated in a previous chapter, transfers must be directly targeted at barriers to equality. Unfortunately, the greatest barrier today in many countries where poverty is endemic is unchallengeable power claimed by despotic rulers. Transfers must be structured to undercut these tyrants and help individuals gain control over their own lives. The nation-state with the nationalism, despots, armies and the enemies it invokes is the chief impediment to a safe and equal world.

Secondly, we must be able to protect ourselves from outsiders who may see us as the other, as people not deserving of their acting responsibly. Adherence to our norms must be expected from anonymous outsiders as well as from citizens. Without this expectation, we are not secure.

Towns used to have city walls with gates that could be shut at dusk to separate those who were known and trusted from those who were not. Such isolation is not possible today, nor is it a good idea as such walls help to create othering. However, the Dominator becomes more dangerous when granted the anonymity of being removed from his group and such a stranger might enter our space, harm us and hope to flee. How can we attain security against such a risk? Our first defence will occur naturally when we are a vibrant, cohesive community. When public spaces are alive with people who feel connected to their neighbours, when those people have agreed that certain acts will be named and countered, and when they are confident that everyone will join them in challenging any offending act, then we will all know that anonymous antisocial acts will be sure to be called out. This makes them less likely. The most effective way to reduce all crime is to assure that there are "eyes on the street," as Jane Jacobs so pointedly noted in her critique of large projects in New York City in her seminal work, *The Death and Life of Great American Cities*.[1] We will be helping to assure the safety of our community when we walk the dog, sit in the café, shop in the town core, meet friends in the park or occupy the windows, doors, porches and decks where our private lives intersect with the public sphere. A connected community with vibrant public spaces and an agreed code of behaviour naturally creates a safe space. Intruders with evil intent will sense this and take their search for victims elsewhere.

Even though the risk may be greatly reduced by this, it is still possible for someone to sneak into our community to rob, assault, vandalize, rape or murder, and then flee. If this fear is substantive enough to require reassurance, then we must bring these concerns to a town meeting. Options are available. The local bobby could organize a group of volunteers to watch the street at times of concern. A whistle, a siren, or a connection to a group of mobile phones could quickly populate the streets at any hour. We—not a group of empowered professionals—can take responsibility for making ourselves secure. The bobby would be acting on our instructions to help us organize our efforts.

If an act of harm has already occurred, then our bobby could use her connections with other towns and with the tools that connect larger areas to identify, locate and detain the perpetrator. We can then demand that he answer to our community by righting the wrong, by restoration of property loss, by recognizing the fear he has generated and, for instances of rape, murder and maiming, to attempt to atone in some fashion for how drastically he has changed people's lives. It will be between the community and the perpetrator to negotiate a way forward. These negotiations would start with the fact that the perpetrator is held by us until agreement is reached. If he wants to reduce his punishment and regain his freedom he must attempt to make amends. He can ask for forgiveness. He can restore losses if the loss is property. He can try to reduce fears by creating assurances that the crime will not happen again.

All this will be difficult for an individual to accomplish. But if the perpetrator is a member of some other family, neighbourhood and town and if his people wish to reclaim him to their community, then his community may choose to stand with him in these negotiations. The many members of a community could more realistically provide restitution. They could also allay our fears by offering to monitor the perpetrator within his own town, at their expense. They could offer a plan to address the problems that may be lying at the root of his aggression. Their goal would be to reintegrate him into his community. Such negotiations, giving standing to the perpetrator, the victims and both affected communities, would have a better chance of success.

If the perpetrator has no community willing to stand with him, either because he has exhausted their patience, been banished as not worthy of reclamation, or has chosen to abandon them, then he would stand alone in these negotiations. It will probably be beyond his ability to provide either restitution or assurance. Only two possibilities are then available. Either the victims embrace forgiveness and let him walk away out of their lives with a promise never to return or they contract with a tool to provide incarceration. The nature of this incarceration would then be the subject of the negotiation. The community needs to feel safe, but they also want to reduce the cost of keeping him locked away forever. The perpetrator would like to negotiate restrictions that were as light as possible. His argument would be that lighter controls permit more ability to support himself leading to a lighter financial burden for them.

A range of options would probably be available for consideration. House arrest with an ankle bracelet that would allow his position to be tracked by a tool may provide assurance while still allowing him to contribute to both his own upkeep and to restitution. A low-security facility, allowing the perpetrator to earn while still providing active monitoring, would give assurance that there will be no repeat, but this would come at greater cost to the victim and her community. Even where maximally secure incarceration is seen as the only option, a route for the perpetrator to earn his way back to a less restrictive position could still benefit both parties. A range of tools would offer a range of solutions for their consideration.

Anger may fuel a desire for revenge, especially in cases, such as rape, murder, assault and maiming, where effective restitution can never be possible. A desire to punish more harshly than is required to guarantee safety will be the expression of this anger. When a perpetrator has his community standing with him, emphasizing those aspects of his life as a son, as a brother, as a father, as a friend and as a worker, he may be seen as someone worthy of another chance and the anger may recede. Standing alone, he is defined solely by one heinous act. He is the monster who has harmed us, easily the focus of our hate.

The dilemma faced by the victims is that they will be further harmed if they indulge their urge to seek revenge. To be responsible agents in the world, we must stand behind every action we take. However justified it may seem in our moment of pain, the effects of a harsh response will be part of our lives forever. Consequences, unseen as we enforce our decision, will land at our feet. If we were to demand capital punishment, we will always be a murderer. Considering this possibility before anger clouded our minds, our town may have chosen to ban such judicial killings, instead providing predetermined guidance for these situations that aim us towards rehabilitation. But anger will have to be recognized as one factor in the negotiations, and that anger is a personal connection between victim and perpetrator. There will be no distant branch of the penal system designated to carry out revenge in our name. It is us, them, our anger, their actions, our need for closure, their fate as an outcast that will define this situation. These negotiations will be difficult.

There may be a need for a long-term incarceration tool whose main promise is to keep individuals who are a proven threat to society isolated for the rest of their lives. Any renegotiation of this sentence to a less restrictive regime would need approval from the victims and this may never be forthcoming. There are actions that can put one outside of society with no guarantee of a way back in—exile has always been a possibility for those unable or unwilling to adhere to the norms established by a community. There would be no expectation of rehabilitation. The purpose would be solely to save society from further damage from these individuals. The cost of this service would continue to exist as a spur for the victims' communities to consider less restrictive situations; the possibility of greater freedom would be the spur to the perpetrator to seek rehabilitation.

As equality increases and poverty disappears, fewer people will emerge from childhood bearing scars from childhood trauma. Eliminating these roots of alienation is the best way to take crimes of violence out of our lives. When bullying will always be named and opposed, then these acts cannot become habitual. Instead, such acts will point individuals towards help in dealing with their demons. But we will have to be vigilant. The Dominator will always be lurking, forever capable of driving us back into that dog-eat-dog world we will have only recently escaped.

CHAPTER SIXTEEN

Conclusions: Eleven Things We Know

∿→

1. Human beings contain both the Cooperator and the Dominator.

Throughout our evolutionary adventure, volcanoes, earthquakes, meteor strikes, ice ages, changing climate, floods, fire and drought have regularly made every human survival strategy ineffective, sometimes gradually, sometimes with stunning swiftness. Only those peoples who could alter their strategy to fit the incoming situation would survive. In those periods of change, knowledge gave us options. We became innovators, inventors and explorers striving to increase our understanding of our world even in times of stability because we knew that knowledge had translated into options in the past. Variety, flexibility and curiosity became the great human strengths.

In every phase, some activities were so crucial to our survival that they became internalized as instinctive drives—thinking about how to react to a predator is less effective than jumping first and thinking later. The instincts necessary to survive in each new circumstance in which we found ourselves differed because the food sources and dangers that conditioned our behaviour had changed. Occasionally, entirely new instincts were inculcated—those who possessed them in some measure thrived; those who did not perished. However, though there was this selection process for developing new urges, there was no evolutionary mechanism to necessarily make old ones disappear when they were no longer relevant. They could persist, urging us to behaviours that could be troublesome in new situations. We developed many urges, each capable of being triggered by circumstances in the world around us.

We dealt with this complexity by creating traditions, taboos and customs to prioritize one drive over others in any given situation. This context of rules determines which aspects of humanity will be permitted to shine and which will be suppressed. Like chameleons, we take our nature from our background.

This range of human character on display has made the essential nature of man a challenge to describe. Some, following the line of Jean Jacques Rousseau, who had seen effective groups where domination was discouraged, argued that we are basically good, though often corrupted by circumstance. Others, following Thomas Hobbes, based on his first-hand knowledge of the English Civil War, take a more pessimistic view, seeing people as needing to be strongly controlled lest our baser instincts prompt us to run amok. The evidence for both is plentiful. We can be either, depending on the setting we construct.

These drives from our evolutionary past, each crucial to our survival in a particular period, are powerful motivators in every human being today. They are the prime initiators of human activity.

16. Conclusions: Eleven Things We Know

Our ancient ancestors were solitary creatures living on a food chain, hunting others while trying not to become dinner themselves. Each competed to survive. The slow, the weak and the stupid did not last long. The strong took what they needed and defended what they had against others who would take it from them. They thrived only if they could become instinctive Dominators.

But life on a food chain was perilous. Many species found their chances of survival to be enhanced when their efforts were combined with those of a group of allies. Acting as packs, herds or troops, they were better at defending, hunting and sharing in times of dearth. Our mammal and primate ancestors have always been part of a group. Over millions of generations, the need for inclusion became a very strong urge. We developed an instinct to coalesce.

This group became our helpers; those outside remained our enemies. The world divided into us and them—our hunting pack, our grooming buddies, our troop versus everyone outside of this extension of ourselves. These others should be dominated, hunted, killed and driven away while our mates were protected, fed, nurtured and treasured. The Dominator still had an arena, but the Cooperator had found a space as well.

When evolution generated the apes, the groups that formed included individuals who differed significantly in strength. The stronger were able to subdue their mates as well as their enemies. Domination crept into the group and a hierarchy resulted. Every member lived in thrall to an alpha male. Their urge to dominate or their fear of domination ordered their lives. That structure defined us for millions of years. It left a powerful mark.

A couple of hundred thousand to a few million years ago, individuals of middling strength were able to form coalitions and successfully oppose these alpha males. When the coalition grew to include the whole troop, taboos and customs that denied domination became an essential part of the structure of their lives. The troop, not the alpha male, once again became our defence against predators. The troop fed us through hunting parties and joint gathering. The troop protected us from others by organizing our efforts in war. Individuals were bereft and helpless without their group. The old urge to cooperate that underpinned this activity became central to our lives. However, though suppressed by taboos, the urge to dominate did not disappear. We still feel this drive. We can still take satisfaction from winning, from being seen as better than others, from taking from those who are weaker. When taboos against this type of behaviour are relaxed, as occurred during feudalism, humans can readily re-adapt to a dog-eat-dog world.

Along with our urge to dominate and our urge to cooperate, we have developed a third drive which pushes us to innovate. Every time that our strategy was thrown into disarray, we had to adapt or die. We were often saved by marginal knowledge and skills that had been uncovered in our past iteration. The fruits of some distant ancestors who had been following their curiosity gave us options. With every success, this drive to discover became stronger. It is another essential part of our nature. A push to understand, to know, to learn, to explore is part of being human.

One of these three drives lies behind every human act. Since these urges can be contradictory, we must rely on traditions, taboos and customs to create a setting that will encourage some activities while denying others. When we construct our societies, we choose which aspect of ourselves will be significant. This social world has become ever more complex as our knowledge, our tools and our possibilities have continued to multiply.

Troops evolved into tribes. As these tribes were wildly successful, their numbers grew. They migrated across the globe. New habitats demanded new survival strategies from our bag of tricks. We adapted and thrived everywhere. Human societies became extremely varied.

2. Feudalism became the structure of our world.

During our tribal period, the taboo against internal violence was paramount. Traditions gave priority to equality and cooperation in every situation. Anyone following their urge to dominate was punished. Innovation was only allowed in severely circumscribed ways. Life continued like this for tens of thousands of years. And then the climate changed. The ice sheets retreated. Traditional foods disappeared. Starvation stalked every hearth. The taboos became unenforceable. In the desperate search for a new stability, a few tribes managed to survive by turning to sowing, reaping and storing crops rather than relying on gathering. They fed, protected and slaughtered animals instead of relying on hunting.

This strategy worked but it raised new questions. The tribe was forced to give up their nomadic life because survival now depended on fixed fields, barns, herds and granaries. Being able to defend these assets was essential, so a group of warriors were charged with defending the granary. Their task in the division of labour was to practice the arts of war while the others became farmers. They became more powerful than the rest of the group. Equality had been sacrificed in the name of defense. Coalitions to depose troublesome alphas were no longer possible. The Dominator had escaped his cage. Feudalism was the result.

There were other tribes who had adopted this strategy. They also had valuable granaries. These fell into the traditional role of "others" so the soldiers, powerful and answering to their greed, were motivated to attack. When they were successful, their fields grew and the tribe could support more farmers and larger armies. Bureaucrats were supported to organize both the farming and the army. Farming villages grew into cities. Demesnes became kingdoms which ballooned into empires. The urge to empire never ceased. They always over-reached themselves, the incipient empire collapsing in devastating war, chaos and famine, leaving fertile ground for the next Dominator. Growth—of fields and territories, of population, of armies, of cities, of accumulated wealth—became the aim and the heart of feudalism. Driven by the urge to dominate and hoard, unchecked by any taboos, the cycles of growth and collapse were unstoppable. This defines our history during the last twelve thousand years.

These soldiers became the rulers of their world, constantly bending everything to their will. They demanded that the human urge to build, to explore and to innovate serve their needs. Their search for better weapons expanded human knowledge of metals, mining and metallurgy. Their need for more impregnable forts led to bigger and more secure architecture, bigger and more secure towns. Their striving for comfort led to advances in clothing, food production and mobility. Alongside the cycles of empire building and collapse, this second theme was also pulsing steadily throughout our history—an inexorable advance in the range and power of our technology.

Cities became areas of wealth generation. The innovation this spawned was controlled by the elites with any new possibilities that might threaten their hegemony summarily squelched. However, in western Europe, significant wealth slipped out of their direct control. Commercial guilds grew powerful enough to successfully challenge the soldiering elite. Once free of aristocratic domination, the wealthy merchants competed amongst themselves, driving innovation beyond weapons, fortifications and basic comforts, and spurring a rapid increase in production, trade and wealth. This competition amongst merchants pushed their societies into exploration, colonialism and industrialization. These merchants were driven by the urge to dominate and hoard every bit as much as the soldiers they supplanted.

Growth was their ambition. As they made more things, they could sell more and accumulate more wealth. This was their route to power and status. Wealth created meaning and a sense of self-worth. An unquenchable thirst for power generated a flow of new products to sell, new markets to buy them, and new technologies to make the production more efficient. This drive to increase production cemented growth at the heart of our world.

Many good things issued from this growth. We now have a prodigious ability to produce and distribute goods: most of us have enough food and lives of comfort. Mechanical devices have replaced much of the backbreaking work that wore down earlier generations. The benefits of hot water on demand, central heating, safe disposal of waste and supermarkets stocked with food are taken for granted in much of the world. Ease of mobility makes every corner of the world accessible to everyone. Famine has virtually disappeared for the first time in human history. Innovations in medicine have transformed many fatal diseases into episodes of illness in long healthy lives. Trade links have made wars more regional and less frequent. Basing our societies on this competitive drive to increase production has delivered many benefits.

As firms grew, one man could no longer control them. Further growth demanded that they be turned into machines defined by procedures, flow charts for materials and information, and detailed job descriptions, all with the singular aim of allowing production to increase. These machines developed programmed connections between themselves. We must protect this machine or risk losing the benefits it provides. Under this industrial machine, growth became the defining feature of our abundant world.

3. Such uncontrolled growth is a problem.

We live in a finite world. The number of trees seemed uncountable at one time, but we are now at a point where they could all be cut down in our lifetimes if the machine demands them as input. The seas were once a vast constant in our lives but we have imperilled even them with our waste and our rapacious fishing habits. The skies above us, once our measure of infinity, have been altered by our waste in ways that threaten the future habitability of the planet. However, none of these realities generate the feedback that will cause the changes needed to alter the destructive processes. Those feedback channels do not exist. They have been pruned in the name of efficiency. The growth continues unchallenged, driven by the rules and procedures encoded in the machine. Even obvious damage and a sober recognition of our physical limits cannot trigger an easing of growth.

When we empowered the soldiers to protect our granaries, we set in motion a process that demands growth. While we were still few, this growth allowed humans to thrive and to increase incredibly in numbers. It spurred magnificent technological marvels like the airplane, cataract day surgery and the Internet. But it did not produce a survival strategy that would make us secure generation after generation after generation into the unseeable future. Constant growth is a relentless denial of this type of stability. Every success only further increases the pace of growth. But unstoppable growth in a finite world is suicidal. This is the flaw in feudalism as a survival strategy. This inherent flaw took ten thousand years to make its presence felt but it has now become obvious. It is time for a change.

4. When growth stops, the problem is different but no more palatable.

The push to more and bigger and more complex things that continually redefines our world does not occur because we need more stuff or bigger stuff or more complex stuff. Rather, it results from competition amongst producers searching for opportunities to invest as they strive to become dominant by accumulating wealth. As the producers morphed from men into machines, this need to invest became mechanistic and even further removed from human control. It only eases when goods can no longer be sold and profits can no longer be made to materialize.

When markets do become glutted, competition intensifies. Prices are lowered. Savings are wrung out of the production process. Workers are forced to work harder for a smaller wage. Expenses are offloaded into the commons to lend support to the bottom line. Weaker businesses fail. Their workers are dumped into poverty, no longer able to buy things or pay their taxes. Sales and entitlements drop further. More businesses fail. Some services cease. Redundant workers further constrict the market. More failures. More poverty. The spiral can only pick up speed.

This road can only end in collapse, despair, destitution, starvation, violence and tyranny.

We have flirted with this scenario a few times in our industrial adventure only to be saved each time by an innovation significant enough to redirect massive growth into a completely new area. But needing to regularly reinvent ourselves like this leaves us open to that instance where it does not happen on cue. One such failure will be all it will take. It also makes the whole system bigger and grow ever faster into totally unknown areas with unstudied dangers.

Real limits imposed by the extent of world resources and the abilities of our world to absorb waste do exist. The unstoppable need to invest is designed to ignore these limits. Causing irreparable harm is intrinsic to the machine. So growth will stop only if markets are glutted, which is a disaster, or if the out-of-control machine transgresses natural limits, which is also catastrophic. A stop to growth is not part of any plan. The exact timing may be unknown, but unfortunately the ends of the process can be seen.

The number of humans on the planet will soon peak and then start to decrease. With fewer people, we will need fewer houses, cars and appliances, less food, furniture and travel, less clothing, less of everything. All markets will be shrinking. This will call for the producing part of our economy to shrink also, but it is not built to do this. Our dilemma will be front and centre.

Fortunately, there is a path out of this conundrum. We are the Cooperator as well as the Dominator. We can construct the beliefs, traditions, taboos, laws and structures that favour that aspect of ourselves. We can value people over serving the machine. We have done so in the past. We can do so again.

5. We control this machine.

It may appear that the machine is being driven by elites who make decisions about investments, production levels, advertising strategies and employee treatment. If that were true, it would provide someone to blame and someone to petition for change. However, these people are just fulfilling scripts over which they have limited options. It is not people, but procedures and structure that define the operations of the machine. These drive us towards maximum growth and deflect activity that interferes with that goal. The job of an executive is to manage growth. Any other goal is out of the question. They are well paid cogs in the machine, but they cannot help us.

We, as workers, consumers and citizens do our jobs, generate demand by choosing amongst products on offer, and combine to create civic tools. These actions all affect the machine. We act and the machine reacts. Growth follows naturally because we adhere to a belief system that asks us to demand more things, larger bank accounts, newer cars, bigger houses, more money, better jobs and the longer hours of work it takes to get these things. When our ancestors were trying to survive with inadequate food, shelter and comfort, these goals were obvious. But we have advanced beyond subsistence and our appetites show no signs of abating. We acquiesce reflexively to every incentive that locks us onto the treadmill of having, wanting and working. We are addicted to the satisfactions of hoarding. We propel the machine forward at an ever more furious pace.

There is no agency that forces us to value a higher wage or a cheaper price ahead of other goals such as clean air, clean water, absence of poverty or the leisure to create. No one forces us to accept that we must become isolated from our community or that we must be disempowered by passive citizenship. In the past, soldiers with swords, spears or guns made sure that no one deviated from their assigned roles. That is not true today. We acquiesce because the machine asks us to. No one makes us do this. Our actions are a choice.

6. Only cooperation can control growth.

We have another option. We surmounted crises in the past by using our ability to coalesce into effective groups, thriving because we were able to act together. This stance is part of being human: we feel impelled to combine our individual lives into a group on which we can rely for safety, aid, meaning, history and the ability to accomplish activities impossible for individuals. During the last ten thousand years, this urge has been opposed by feudal structures designed only to turn us into more efficient labourers. These structures persist but no one enforces them. They have power only when we willingly embrace them. We can choose to act differently. A communal society, also natural to us, will provide satisfactions different from those that flow from hoarding and domination. And when we express communal values in our lives, growth loses its destructive power.

We can choose to create a family. We are free to give a value to equality, empowerment and responsibility that exceeds the value we give to hoarding and cheapness. These values will define our world as we express them in our lives. This will change our world.

We need a family that is small enough for everyone to sit around a table able to make eye contact but large enough to include variety in abilities, personalities and coalitions. This group is strengthened when it joins with other families to become a neighbourhood and these neighbourhoods are enhanced when they coalesce into towns. Each level of cooperative activity increases the actions done as gifts and so reduces the thrust for growth.

In towns, where each person is known by sight and reputation, an organization based on trust can emerge that is capable of focussing the efforts of thousands on the particulars of their lives. Much that is now ordered only by competition will be replaced by activity arising from cooperation. Growth will disappear. The choice is ours to make.

Coalitions of towns can build tools that meet the needs of tens of thousands or millions of individuals. A tool designed so that control rests with the user can assure that activities meeting the needs of huge numbers of people are still defined by responsibility, a desire for equality and cooperation.

When we are no longer driven to compete against our fellows to survive, we can aim towards any goal we want. We can work less, live well and heal the world. It does not have to be one of these at the expense of the others. Our future will be limited only by our ability to dream.

This powerful urge to coalesce into groups, though potentially our saviour, does contain a danger. Strong group ties create solidarity, but at a cost of hostility towards those outside this group. Throughout feudal history, this impulse to "other" has been exploited by elites to mobilize support for their territorial ambitions. As our families strengthen, we must stay open to those outside our groups. Failure to embrace the "other" has been the Achilles heel of all attempts at a cooperative life.

The good news here is that humans are malleable. We have shown through the loyalties we easily give to large states that instinctive hostility to others can be softened by a recognition of the things we share. Information about others, connections of trade and travel, common symbols, common institutions and a genuine desire to create equality by sharing resources can replace hostility with brotherhood. We can choose to create a family of the eight billion members of humanity.

We can create successful human-scale groups. It will be economically advantageous. It promises to be psychologically satisfying. It can replace isolation with meaning. Basing our lives on such groups will undercut the impetus to growth that is driving our world into desperate straits.

7. True security can exist.

When our security relies on accumulating assets, we must be constantly competing. But when our security rests on the assured support of a coalition of equals, we will be able to pursue other goals such as clean air, clean water, work/life balance, beauty and fulfilment. We can escape the passive acceptance of the bad as the cost of the good. Choosing the assured sustenance of human-scale groups is key to a stable future.

Basing security on a nest egg is illusory. Throughout history, assets have been wiped out by natural disasters, economic crashes and war. In these disasters, the existence of a caring family and friends always spelled the difference between those who failed and those who survived. Robust families, neighbourhoods and communities are what really provides security.

8. Equality is key.

A coalition dedicated to seeking equality while naming and opposing domination needs everyone. All must stand together. All must know that their contribution is essential. Within any group, there cannot be some who are takers and others who are givers, some giving orders while others carry them out. When we share food, risks, tasks and meaning, we make equality a fact and not just a goal. When authority is earned through reputation and not taken with power, we will have organization with responsibility. When equality is weakened, cohesion retreats, group effectiveness dissipates, and competition and hoarding creep back in. We must value equality and act to make it real.

This is true in every group of which we are members—family, neighbourhood, town, region and the entire globe. As equality grows in each sphere, envy, fear and conflict retreat, liberating our energies to create.

9. Sharing work is the magic bullet.

The competition for jobs and wages is a major driver of growth in our economy. When we gain a more prestigious job or get a raise in salary, we display to the world that we are better than others. We get to claim the esteem due to one who sits high in the machine. This satisfies both the Dominator and the Hoarder in us. Chasing this esteem drives many lives. This race is fuel for the machine, driving growth to new heights.

When we are secure in our family, neighbourhood and town, we no longer need the vision of the next promotion to be the driver of our lives. Instead, putting time and effort into strengthening our human-scale groups becomes more important. These groups are healthiest when everyone contributes to tasks as well as earning a wage.

With the productivity of a modern economy, there is a decreasing number of hours of work that needs to be done. We can choose to share this work. It is up to us. This will involve people giving up shifts or contracts or hours so that underemployed people can have wage work. Underemployment and unemployment will be signs that everyone must seek ways to reduce their hours. This follows from prioritizing equality over amassed wealth.

This sharing cannot be enforced from above. That re-introduces the seeds of tyranny. It must arise from personal choices taken from a desire to bolster equality, embraced as the route to de-throne growth from the centre of our world. Sharing our jobs is where the rubber meets the road.

Sharing jobs will reduce everyone's wage work contribution to around one or two days a week. The recaptured time will enhance our lives, our families, our neighbourhoods and our towns. How we spend this time creates the world.

10. Transfers are the force that unites the world.

Responsible people stand behind their activities, no matter how distant the effects in time and space. Global inequities are a residue of our feudal past. Some trace their roots to geographical differences in climate and richness in sources of food. Some arose from advantages conferred by differing rates of technological development. Some were just accidents of history that nudged development onto a different path. But as feudalism matured, these differences gave an advantage to the wealthy in their quest to exploit. They plundered the less powerful regions for cheap labour and cheap resources to meet our demand for cheapness. This exploitation continues. We continue to benefit. We are responsible.

Transfers can heal this wound. They are a tangible expression of a desire for equality. They embody a wish that the world become a coalition of strong equal regions. Sharing our jobs within our community and around the world is a transfer of opportunity. Today, many people are forced into inactivity or an inefficient scrabbling for survival whereas sharing our work will allow all to contribute. That act will liberate ourselves while empowering others. Strong equal regions will allow everyone in the world to collaborate in attacking the existential problems bequeathed to us by our feudal past.

The equal world that meaningful transfers will deliver has transformative benefits for everyone. War can be eliminated. Hunger can be banished. Armies can be disbanded and the waste of military spending can be redirected into real benefits. Nuclear stockpiles can be dismantled and turned into abundant energy. We can all cooperate in neutralizing the abomination of chemical weapons. Habitats that have become degraded can be restored without border protection and rapacious businesses thwarting our attempts. Global problems that require a global citizenry can be addressed. Transfers can create that unity.

Valuing equality is a choice. Embracing transfers is a choice. This sunny future is a choice.

11. We can do it.

Historical golden ages arose in the past when human-scale groups were strong and equality was valued. Time for human beings to follow their curiosity in a world where enhancing the community was important led to innovation, expression, creation and exploration. Identity with community sparked both public art and philanthropy. Living in a vibrant community produced security, safety and meaning. It has happened before. It can happen again.

Basing our future on our urge to cooperate and our need to be embedded in a nurturing community is natural to human beings. The actions that we need to usher in this golden age are natural to us.

The ten-thousand-year experiment of basing the organization of our world on an unfettered urge to dominate is ending. Time has exposed its inherent contradiction. The threat is real. We have an alternative.

During this same ten-thousand-year interval, another aspect of humanity has also been maturing. The Innovator has continued to expand human understanding. We now have the ability to anticipate approaching difficulties and change before the flaw forces collapse. No other animal with a flawed survival strategy has been able to take a hand in choosing their future. We are responsible for our future. The next phase of our history is up to us.

Notes

Chapter 1: Beginnings

1. "Big Bang Theory," Encyclopedia.com, May 23, 2018, https://www.encyclopedia.com/science-and-technology/astronomy-and-space-exploration/astronomy-general/big-bang-theory

2. Broadly, see Michael Allaby, Dictionary of Ecology (Oxford University Press).

3. Fred Hoyle, *Ice: How the next ice age will come – and how we can prevent it* (London: Hutchinson, 1981), 138-141.

4. Loren Eiseley, *The Immense Journey: An Imaginative Naturalist Explores the Mysteries of Man and Nature* (New York: Modern Library, 1957), 54.

5. Broadly, see Elaine Morgan, *The Aquatic Ape* (New York: Stein and Day Pub., 1982).

6. "East African Rift," Wikipedia.org, accessed April 13, 2023, https://en.wikipedia.org/wiki/East_African_Rift

7. Tristan Free, "Human fat storage: how we became the 'fat primate'" BioTechniques.com, July 4, 2019, https://www.biotechniques.com/cell-and-tissue-biology/human-fat-storage-how-we-became-the-fat-primate/

8. David Pilbeam, *The Evolution of Man* (London: Thames & Hudson, 1970), 89.

9. Matt Ridley, *How Innovation Works* (New York: Harper, 2020), 223-224.

10. Elaine Morgan, *Descent of Woman* (New York: Stein and Day Pub., 1972), 125-126.

11. Elaine Morgan, *The Scars of Evolution: What Our Bodies Tell Us About Human Evolution* (Oxford: Oxford University Press, 1994), 169.

Chapter 2: The Originator

1. Jared Diamond, *The Third Chimpanzee: The Evolution and Future of the Human Animal* (New York: Harper Perennial, 1992), 41-42.
2. David Pilbeam, *The Evolution of Man* (London: Thames & Hudson, 1970), 45, 204.
3. Broadly, Richard Wrangham, *Catching Fire: How Cooking Made Us Human* (New York: Basic Books, 2009).

Chapter 3: The Originator Breaks Free

1. Robert Lowie, *Indians of the Plains/Indians of the Northwest Coast* (American Museum Science Books, 1963), 124-125.
2. Marvin Harris, *Cannibals and Kings: Origins of Cultures* (New York: Random House, 1977), 13.
3. Ibid., 10-11.
4. Jane Jacobs, *The Economy of Cities* (New York: Vintage Books, 1970), 18.
5. Andrew Sherratt, "The Obsidian Trade in the Near East, 14,000 to 6,500 BC," ArchAtlas.org, 2005, https://www.archatlas.org/journal/asherratt/obsidianroutes/

Chapter 4: The Growth of the State

1. William McNeil, *The Rise of the West: A History of the Human Community* (University of Chicago Press, 1963), 198.
2. H. D. F. Kitto, *The Greeks* (London: Penguin, 1973), 158-160.
3. Paul Goodman, *People or Personnel: Decentralizing and the Mixed System* (New York: Random House, 1965), 151-152.
4. Lauro Martines, *Power and Imagination* (New York: Vintage Books, 1980), 257.

Chapter 5: The Inevitability of Empire

1. Harry Magdoff, *Imperialism: From the Colonial Age to the Present* (New York: Monthly Review Press, 1978), 103.
2. Felix Greene, *The Enemy: What Every American Should Know About Imperialism* (New York: Random House, 1970), 53-54.
3. Arno Mayer, *The Persistence of the Old Regime: Europe to the Great War* (New York: Pantheon Books, 1981), 305.
4. Ibid., 321-322.

Chapter 6: The American Empire

1. Harry Magdoff, *Imperialism: From the Colonial Age to the Present* (New York: Monthly Review Press, 1978), 80.
2. Paul Goodman, *People or Personnel: Decentralizing and the Mixed System* (New York: Random House, 1965), 32.
3. C. Wright Mills, *The Power Elite* (Oxford University Press, 1956), quoted in Felix Greene, *The Enemy: What Every American Should Know About Imperialism* (New York: Random House, 1970), 73.
4. J. A. Hobson, *Imperialism: A Study* (New York: Cosimo Classics, 2005), quoted in Felix Greene, *The Enemy: What Every American Should Know About Imperialism* (New York: Random House, 1970), 107.
5. S. D. Butler, "America's Armed Forces: 2. 'In Time of Peace': The Army", *Common Sense*, 4, no. 11 (1935).
6. A. J. P. Taylor, *The Origins of the Second World War* (New York: Simon & Schuster, 1996), viii.

Chapter 7: Life in the Machine

1. "Mutual Fund," Wikipedia.org, accessed July 3, 2023, https://en.wikipedia.org/wiki/Mutual_fund
2. Michael Brough and Ashvinder Kaur, "International Pension Plan Survey 2023," Wtwco.com, February 28, 2023, https://www.wtwco.com/en-us/insights/2023/02/international-pension-plan-survey-2023
3. David Foster Wallace, 2005, "This is Water: Some thoughts, Delivered on a Significant Occasion, on Living a Compassionate Life," transcript delivered at Kenyon College, May 21, 2005.

4. "Total Fertility Rate," Wikipedia.org, accessed September 18, 2020, https://en.wikipedia.org/wiki/Total_fertility_rate
5. Darrell Bricker and John Ibbitsen, *Empty Planet: The Shock of Global Population Decline* (Toronto: McClelland & Stewart, 2019), 46.
6. "'Population bomb' might prove to be a dud: Study," *Toronto Sun*, April 2, 2023, https://torontosun.com/news/world/population-bomb-might-prove-to-be-a-dud-study
7. "Don't Panic," *The Economist*, September 24, 2014, quoted in Darrell Bricker and John Ibbitsen, *Empty Planet: The Shock of Global Population*, 46.
8. Ana Gonzales Barrera, "Before COVID-19, more Mexicans came to the U.S. than left for Mexico for the first time in years," Pew Research Centre, July 9, 2021, https://www.pewresearch.org/short-reads/2021/07/09/before-covid-19-more-mexicans-came-to-the-u-s-than-left-for-mexico-for-the-first-time-in-years/
9. Daisuke Hori, "More empty homes: Japan's housing glut to hit 10m in 2023," *Nikkei Asia*, September 11, 2022, https://asia.nikkei.com/Spotlight/Datawatch/More-empty-homes-Japan-s-housing-glut-to-hit-10m-in-2023.

Chapter 8: Taking Control

1. Max Roser, "Why did renewables become so cheap so fast?" Our World in Data, December 1, 2020, https://ourworldindata.org/cheap-renewables-growth
2. "Fusion power is coming back into fashion—This time it might even work" *The Economist*, March 22, 2023, https://www.economist.com/science-and-technology/2023/03/22/fusion-power-is-coming-back-into-fashion
3. "Progress and Poverty," Wikipedia.org, accessed April 3, 2023, https://en.wikipedia.org/wiki/Progress_and_Poverty
4. Broadly, Gabor Maté and Daniel Maté, *The Myth of Normal: Trauma, Illness, and Healing in a Toxic Culture* (Toronto: Knopf Canada, 2022).
5. Rutger Bregman, *Utopia for Realists: How We Can Build the Ideal World* (New York: Back Bay Books, 2018), 25.

Chapter 9: Cooperation

1. Jay Walljasper, "Elinor Ostrom's Eight Principles for Managing a Commons," *On The Commons*, Oct 2, 2011 https://dlc.dlib.indiana.edu/dlc/bitstream/handle/10535/7708/Elinor%20Ostrom's%208%20Principles%20for%20Managing%20A%20Commmons%20_%20On%20the%20Commons.pdf
2. D'Arcy Wentworth Thompson, *On Growth and Form* (Cambridge University Press, 1961), 24.
3. "Dunbar's Number," Wikipedia.org, accessed March 15, 2023, https://en.wikipedia.org/wiki/Dunbar%27s_number

Chapter 10: Organizing Our Cooperation

No notes in this chapter

Chapter 11: Eve

No notes in this chapter

Chapter 12: Work

1. "Bullshit jobs and the yoke of managerial feudalism," *The Economist*, June 29, 2018, https://www.economist.com/open-future/2018/06/29/bullshit-jobs-and-the-yoke-of-managerial-feudalism

Chapter 13: Responsible Infrastructure

1. "What are the safest and cleanest sources of energy?" Our World in Data, accessed May 25, 2023, https://www.ourworldindata.org/safest-sources-of-energy
2. "Traveling Wave Reactor," Wikipedia.org, accessed February 9, 2021, https://en.wikipedia.org/wiki/Traveling_wave_reactor

Chapter 14: Looking After Each Other

1. Yuval Noah Harari, *Homo Deus: A Brief History of Tomorrow* (Toronto: McClelland & Stewart, 2016), 21.
2. Abraham Maslow, "A theory of human motivation," *Psychological Review*, 50, no. 4 (1943), 370-396, https://doi.org/10.1037/h0054346

Chapter 15: The Dominator

1. Jane Jacobs, *The Death and Life of Great American Cities* (New York: Vintage, 1992), 29 and broadly.

Chapter 16: Conclusions: Eleven Things We Know

No notes in this chapter

About the Author

Garry spent 35 years running a bookstore connecting others with the words that would enhance their lives. During that time, the ideas that inhabited the books on his shelves, the experiences of running an endangered business, and his initial training as a mathematician and engineer, all prodded him to connect the many disparate facts of a modern life into an understanding of why human-created systems operate as they do. These ruminations culminated in a broader vision of how the world could work once we got beyond left and right. Now retired, he has distilled those thoughts into this offering.

www.ingramcontent.com/pod-product-compliance
Lightning Source LLC
Chambersburg PA
CBHW031055080526
44587CB00011B/692